A CONCISE
ADVANCED
GEOGRAPHY

TIM BAYLISS

OXFORD UNIVERSITY PRESS

1995

Oxford University Press
Walton Street Oxford OX2 6DP

Oxford   New York
Athens   Auckland   Bangkok   Bombay   Calcutta
Cape Town   Dar es Salaam   Delhi   Florence
Hong Kong   Istanbul   Karachi
Kuala Lumpur   Madras   Madrid   Melbourne
Mexico City   Nairobi   Paris   Singapore
Taipei   Tokyo   Toronto

and associated companies in
Berlin   Ibadan

Oxford is a trade mark of Oxford University Press

© Oxford University Press 1995

ISBN 0 19 914660 8 (Non-Net Edition)
ISBN 0 19 914667 5 (Net Edition)

Printed in Great Britain.

Acknowledgements

The publisher and author would like to thank the Earth Satellite
Corporation/Science Photo Library for permission to reproduce the
front cover satellite image of London.

Artwork by Mike Gunnell.

Acknowledgements must go to those authors who have maintained my
undiminished interest in geography and whose examples have been so
influential in the production of this text. Thanks also go to Ann Wright
for her invaluable comments on the draft, to Mike Gunnell for
redrawing all illustrations, and to Angela Airey for preparing the index.
However, it is dedicated to Sue Chaney and past and present
sixth-formers of Hull High School, without whose influence it would
not have been possible.

PREFACE

The study of geography post-sixteen is without doubt appropriately demanding, intellectually stimulating, and relevant. *'The one true science'* [1] accompanies its students through life – enhancing their understanding and appreciation of infinite environmental and human dimensions of a dynamic, complex world. Popular amongst sixth-formers and appreciated by institutes of continuing education, expanding numbers continue to study a discipline which lends itself particularly well to contemporary 'modular' approaches. Textbook production has met this demand with ever more appropriate publications in often attractively packaged formats. Students enjoy widespread and stimulating choice – certainly not restricted to the *Selected General Reading* offered at the end of this text. However, a complementary, concise, virtually synoptic text, produced economically enough to be purchased outright, would enable students to highlight and annotate further without fear of defacing a resource that others must inherit. The adoption of straightforward line illustrations, with neither photographs, colour, nor any pretence of fully comprehensive discussion and exemplification, addresses this need. Indeed, all the material in this text has been tested on sixth-form students with this objective in mind and amended according to their germane observations. Clearly, in its writing, decisions on content, and especially omissions, have proved extremely difficult. The final decisions on topic weightings, the inclusion of a comprehensive index, text highlights, selected key definitions and brief case studies, the explanation of traditional theories to include appraisals, an *en passant* approach to many development and environmental issues – not least meteorology and climatology – have been influenced strongly by constructive comments from students in a number of institutions. Their perceptive comments on both the content and approach of established textbooks have identified topics and issues covered thoroughly in both depth and summary that less emphasis, beyond, say, contextual references, need be adopted in this text. Clearly, therefore, established textbooks of substance cannot be replaced, for they provide comprehensive coverage of the discipline, with original illustrations. This concise text is intended solely to complement and enrich the study of these and other resources – in a cost-effective format appropriate for coursework reinforcement, ready reference, and examination revision.

[1] Simon Jenkins – former editor of *The Times*.

CONTENTS

INTRODUCTION

A Concise Advanced Geography represents 'student power' in, hopefully, a particularly useful form. Throughout my teaching career I have taught post-sixteen courses and witnessed a considerable variety of reactions to various syllabuses and textbooks. Particularly as a head of department, I have increasingly sought feedback from students, both in my own school and other institutions, on what resources they would find most useful and why. I, like many colleagues, have delighted in the increasingly high standards of textbook production in recent years, assuming that all students would be similarly enthusiastic. However, I have found their sober, practical considerations, often involving stress and worry, to be more prevalent than the enthusiastic subject teacher might realise. Issues such as how to cope with considerable workloads across chosen subjects, increasing personal and parental expectations regarding continuing education, and the inevitable interruptions to course progression due to open-days, interviews, and health absences have been expressed repeatedly. Particularly interestingly, feelings of guilt have been revealed in that many of the texts available are not used to full advantage due to their sheer depth of content and bulk. Other reactions, albeit less common, have been worries regarding the availability of sufficient texts due to their considerable cost and fragility under heavy use. Candid student perspectives can certainly surprise and perhaps inevitably this feedback eventually blended with suggestions from my own students that I should stop looking for the 'affordable, annotative, not too substantial, but certainly not superficial course companion' that so many hanker for, but to get on and write it.

As stated in the Preface, the style and content is strongly student-influenced, and the book is produced economically enough for institutions or students to buy it outright in order that additional annotation or highlighting on text and diagrams may be encouraged. The text, whilst concise, is more substantial than simply an aid to revision. Brief case studies have been included in support of the theory, for both (essay writing) reference and learning - not least to illustrate the style and depth appropriate to answering examination questions at this level. The style of writing adopted may be read at different levels and in different circumstances. Its 'weighting' is pitched in recognition of sixth-formers having to cope, under pressure of time, with three or more subjects at this level. The volume of many standard texts can both intimidate and literally prohibit their use in certain circumstances – such as reinforcement reading whilst travelling. It is intended, therefore, that *A Concise Advanced Geography* should lend itself to repeat and speed reading. Furthermore, the adoption of hand–drawn line illustrations should remind students of their value in communicating information in an examination situation. What is memorable is reproducible.

Clearly, however, any suggestion that this book is sufficient in itself for examination preparation must be refuted. Whilst some reviewers of the trial copy have commented that it is a great comfort as a supporting text, there can be no substitute for reading in depth of established textbooks if examination success at this level is to be achieved. This text's middle ground between 'full-bodied' textbook and revision study aid does seem, however, to have struck a notable chord; many have commented on its flexibility as a reinforcing device in support of conventional texts, which is also invaluable for revision. A principal objective has been that the final product should be cheap enough to purchase outright rather than use with a view to other students inheriting the inevitably 'dog–eared' copy at a later date. The book has also been designed so that text spacing and general layout are appropriate for further highlighting and diagram annotation, but in a package small enough to carry easily and use beyond formal study areas.

The comprehensive contents section and index are intended to speed up access to the text as a ready reference. Italics in the former, for example, indicate a boxed feature such as a listing of key factors, brief case study, technique, or definition not integrated into the text. In the index, a page number in bold indicates a main entry, such as the term being defined or explained in full. Normal print, however, indicates a reference in context, whilst any number in italic, whether bold or not, tells you that the word is used in a boxed feature.

1
EARTH STRUCTURE

The earth is a geoid (ellipsoid of revolution) - marginally flattened at the poles due to centrifugal forces generated by its rotation. Its composition remains the subject of much research and debate, yet is of great relevance to our understanding of **endogenetic (internal) processes** which account for nearly all the earth's major features of relief such as mountain chains, plateaux, basins, and ocean trenches.

The **crust** is effectively the shell or skin. It varies in thickness from around 5 km under oceans to over 50 km under continents. The crust along with the upper mantle is known as the **lithosphere** and is of particular relevance to geographers.

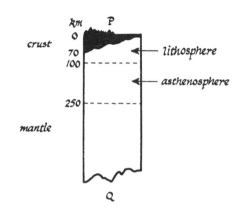

The **mantle** is 2900 km thick. The upper mantle consists of ultrabasic solid rocks on top of a plastic zone known as the **asthenosphere** and capable of slow flowage. Densities progressively increase as you go down into the lower mantle.

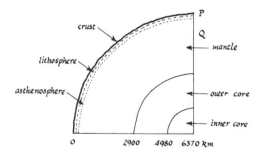

The **core** (solid inner, but liquid outer) has a radius of approximately 3400 km and consists of nickel/iron compounds. The increasing pressure and temperature with depth from the earth's surface reach phenomenal values at this, the centre of the earth.

Sial, sima, and isostasy

The crust is composed of a discontinuous layer of **basaltic** rocks known as **sima**, because they are made up of **si**lica and **ma**gnesium. This used to be regarded as continuous but should now be recognised as interrupted by bodies of **granitic** rocks known as **sial**, because they are made up of **si**lica and **al**uminium, with remnants of sima beneath. These form the continental land masses. Continental crust is predominantly, therefore, sial with some sima below. It is, in consequence, much thicker than the oceanic crust.

Isostasy was long thought to refer to the balance between these basaltic and granitic rocks which are at different densities. Basaltic (oceanic) crust is of a higher density than the granitic crust on top. Each mountain range is balanced by a 'root', sometimes including sima, projecting deep into the mantle. Indeed, the whole crustal system is constantly seeking **isostatic equilibrium** by adjusting to the major forces acting upon it. When glaciation occurs, for example, the weight of ice depresses the continent's sial further into any sima. When the ice melts the continent readjusts by rising in a very slow process called **isostatic recovery**. (The whole process can be likened to pushing a rubber duck under water in a bath - then releasing the pressure!)

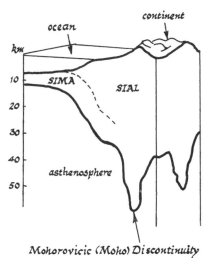

Mohorovicic (Moho) Discontinuity

Prolonged erosion of mountains causes similar isostatic recovery just as deposition of the resulting sediments in lowland basins will depress the sial into any sima. Note the implication here, that movement can take place within the crust. Of most importance, however, is that the whole crust is of lower density than the upper mantle, hence the analogy that it 'floats' – so important to understanding plate tectonics. Isostasy is, therefore, most succinctly described as the crust adjusting up or down relative to the upper mantle.

The **Mohorovicic (Moho) discontinuity** marks the junction between the crust and the upper mantle. Its depth will vary according to the crustal thickness above.

Britain is still recovering its isostatic equilibrium from the last Ice Age. This **isobase map** shows the Isle of Arran rising by over 2 mm per year relative to London sinking by 4 mm per year.

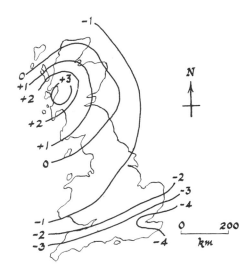

The −2 isobase corresponds to the ice limit in the last glaciation.

Continental drift, palaeomagnetism, sea floor spreading, and plate tectonics

A. Wegener's **theory of continental drift** (1912) suggested that all the present continents were joined together as *Pangaea* before 'drifting' apart into *Laurasia* and *Gondwanaland*. These then broke up to form today's, albeit slowly changing, continental arrangement. The theory was based upon the 'jigsaw fit' of many continents with matching rock types, mineral deposits, fossil plants and animals, and even striations (scratches on rocks) from glacial periods on *Gondwanaland*. However, it was regarded with great scepticism until refined by the work of H.H. Hess (1962) applying studies of palaeomagnetism ('fossil compasses') to sea floor spreading.

Palaeomagnetism studies the history of changes in the earth's magnetic field, which can be likened to a giant bar magnet. Periodically, the magnetic field reverses and any magnetite (iron oxide) formed in cooling magma will record the magnetic orientation of the time. Magnetometers towed behind ships in the Atlantic Ocean recorded a symmetrical pattern of changes in the magnetic field in progressively older rocks away from the Mid-Atlantic Ridge.

The oldest lavas were furthest from the ridge and showed a reversal of magnetic polarity every million years or so. The result is, effectively, a mirrored magnetic bar code.

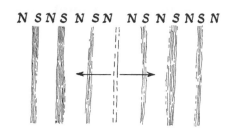

The **theory of plate tectonics** has thus evolved. **Tectonic** means *'a large-scale earth movement'* whilst the **plates** refer to irregularly shaped and sized 'rafts' of both oceanic and continental crust effectively floating on the plastic asthenosphere beneath. The resulting 'jigsaw' is dynamic - the plates moving relative to each other and, we think, driven by vast slow-moving convection currents within the mantle. It is at plate boundaries, of which there are three types, however, that most of the world's major landforms occur, relating to the earthquake, volcanic, and mountain building zones located there.

1. **Constructive (divergent) plate margins** are normally associated with **sea floor spreading**. Molten magma upwells from the mantle to fill the gap and form new basaltic rocks on the sea bed. The spreading takes place in sections separated by **transform** faults. The Mid-Atlantic Ridge, for example, is widening by up to 90 mm a year - a rate equivalent to the growth of our finger nails! **Submarine volcanoes** are associated with these margins and may grow to break the surface and form new islands such as Surtsey, south-west of Iceland - the latter being a rare case of divergence above sea level.

Mid·Atlantic ridge

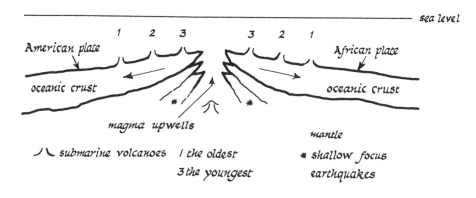

2. **Destructive (convergent) plate margins** are associated with **subduction**. When an oceanic plate collides with a continental one, the former, although more dense, is forced, due to greater mobility, beneath the latter and destroyed by melting at depth. A **deep ocean trench** marks the exact boundary. Long chains of **fold mountains**, such as the Andes, including **volcanoes** associated with rising plumes of melted, hence less dense, oceanic crust, and earthquakes, relating to the increased pressure, are the characteristic continental results. Should two oceanic plates collide, the rising magma forms crescents of submarine volcanoes, which may grow to form **island arcs**, such as the Philippines. Two continental plates colliding leads to particularly high fold mountain formations, such as the Himalayas, because the crust thickens without subduction. Earthquakes occur, but volcanoes are rare because of the extreme crustal thickness.

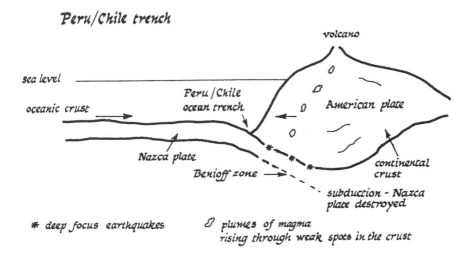

Peru/Chile trench

volcano

sea level

oceanic crust

Peru/Chile ocean trench

American plate

Nazca plate

Benioff zone →

continental crust

subduction - Nazca plate destroyed

* *deep focus earthquakes*   *O plumes of magma rising through weak spots in the crust*

3. **Conservative plate margins** are characterised by transform faults whereby movements are parallel relative to each other. Friction locking the plates together allows stress to build up with high vulnerability to earthquakes as a consequence. The San Andreas fault zone is a classic example.

earthquake zones along transform faults

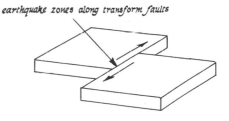

Present plate margins are areas of earthquake and volcanic activity and young fold mountain chains.

The interiors of present plates and the **abyssal plains** (deepest parts) of the oceans are areas of crustal stability.

Past plate margins may now be fused together into stable areas with evidence of ancient volcanoes and old fold mountains such as the Urals.

Just as debate persists over the exact explanation of isostasy (see earlier discussion) so plate tectonic theory is continually being refined. The nature and role of mantle convection currents raise most speculation. Are they fuelled by radioactive processes? Are they initiated by gravitational differences? Are they perpetuated by the earth's rotation? All these factors are likely to be involved. But are these convection currents confined to the relatively thin asthenosphere? Given the size of the plates this is difficult to comprehend - hence research must go on.

World structure

World structure is clearly related to plate tectonic theory and has three key elements:

1. Each ancient continent has a nucleus or shield area called a **craton**. This comprises ancient (Pre-Cambrian) crystalline and metamorphic rocks (see later) which are rich in metals and other minerals.

2. Around the shield edges fold mountains are formed from sediments compressed by 'drifting' continental crust. These are in various stages of denudation according to their age.

FOLD MOUNTAIN EXAMPLES

Caledonian remnants (c. 390 million years old) are found severely eroded in Scotland.

Hercynian (Armorican) folds (c. 275 million years old) are generally planed down or left in blocks such as the Urals, or Vosges in France.

Alpine folding (c. 25 million years ago), such as in the Swiss Alps, is the most recent and so least eroded - with dramatic relief resulting.

3. Most continents also have basins and plains of recent sedimentation such as the North European Plain. Many seas, such as the Mediterranean, which are areas of contemporary sedimentation are, theoretically, the fold mountains of the future.

Earth movements

There are two categories - both are slow, gradual, and of continental proportions. Lateral and vertical movements are best discussed separately from the more rapid, and confined, earthquakes which often accompany them.

1. **Lateral movements** are due to horizontal (tangential) forces generally parallel to the surface. They include plate movements and mountain building (orogenesis) resulting from folding, faulting, and rifting.

Folding is responsible for large orogenic belts such as the Andes and Alps. At smaller scales layers of rock under compression may bend up to form an **anticline** (upfold) which is consequently stretched under tension and so weakened and more vulnerable to erosion. A **syncline** (downfold), by contrast, is squashed under pressure and so is more resistant to erosion. This can have unusual effects on the landscape such as the **inversion of relief** found in Snowdonia, Wales, where mountains such as Snowdon are upstanding synclines adjacent to anticline vales such as Ffestiniog. Similarly, if compression continues, then simple folds become **asymmetrical**, then **overfolded** - even to the extent of 180°, making a **recumbent fold**. Increasing the compression yet further would make the middle section so thin that it might break, creating a **nappe**.

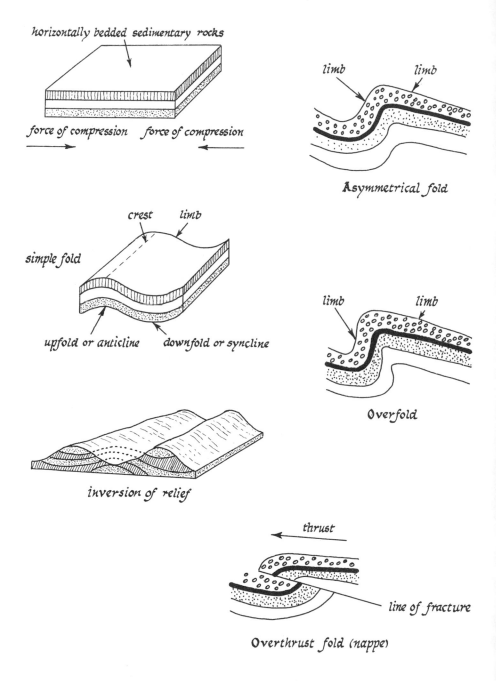

horizontally bedded sedimentary rocks

force of compression   force of compression

limb   limb

Asymmetrical fold

crest   limb

simple fold

upfold or anticline   downfold or syncline

limb   limb

Overfold

inversion of relief

thrust

line of fracture

Overthrust fold (nappe)

Finally, **domes** may result if pressure is exerted from four sides. Again, weakening due to the consequent stretching may result in vulnerability to erosion, such as found in Robin Hood's Bay, North Yorkshire.

Faulting occurs when rocks under the stress of compression crack or slip. Rocks under tension will always fault, rather than fold, whilst rocks under torsion (a twisting action) will tear. There are consequently three types of fault - normal, reverse, and tear.

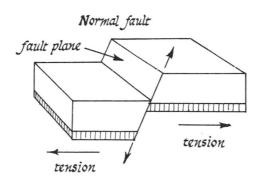

Normal fault

fault plane

tension

tension

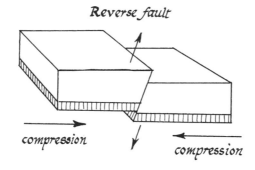

Reverse fault

compression

compression

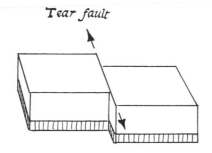

Tear fault

The **throw** of a fault is the amount of displacement - the raised block being the **upthrow**, the fallen block, the **downthrow**. In the extremely rare cases of **reverse faults**, because compression usually leads to folding, the overhanging part of the upthrow is soon eroded.

The Mid–Craven fault in the Yorkshire Dales is a particularly dramatic example of a **normal fault** causing a fault-line scarp.

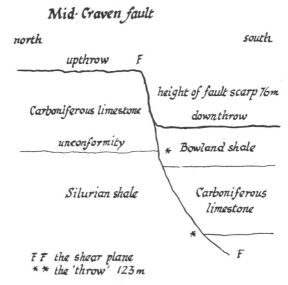

Mid-Craven fault

north

south

upthrow   F

height of fault scarp 76m

Carboniferous limestone

downthrow

unconformity

* Bowland shale

Silurian shale

Carboniferous limestone

*

F

F F the shear plane
* * the 'throw' 123m

(younger rocks eroded from the top of the limestone plateau on the upthrow side)

Above Malham, however, the North Craven fault has little effect on the landscape - but demonstrates geological evidence of the considerable friction and heat which is sometimes involved in faulting. This can **metamorphose** rocks - in this case creating a narrow band where shale and limestone have been turned into slate and calcite respectively.

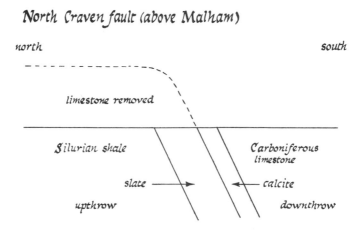

North Craven fault (above Malham)

north

south

limestone removed

Silurian shale

Carboniferous limestone

slate ⟶

⟵ calcite

upthrow

downthrow

Tear faults may produce a spectacular or, conversely, minimal effect on the landscape. For example, in the Vale of Taunton, only soil colour and land-use suggest movement, whereas in Glen More, the Great Glen fault created a **shatterbelt** zone of weakness which was later eroded and enlarged by scouring glaciers.

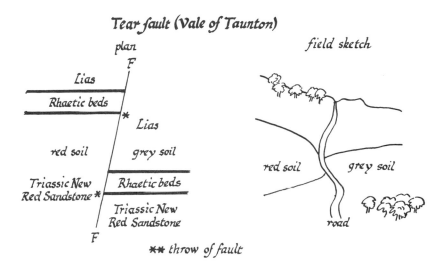

Tear fault (Vale of Taunton)

plan   *field sketch*

F

Lias

Rhaetic beds

* Lias

red soil   grey soil

Triassic New   Rhaetic beds
Red Sandstone *

Triassic New
Red Sandstone

F

red soil   grey soil

road

** *throw of fault*

Rifting also results from tension in the crust. **Rift valleys** form where two sets of faulting have allowed the central area to sink, forming a wide trough-like valley with upstanding blocks on either side called **horsts**. The Rhine Rift Valley is a much quoted example, along with the Great African Rift Valley system which is so large and unique that it almost certainly results from diverging tectonic plates.

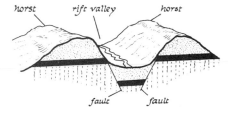

horst   rift valley   horst

fault   fault

2. **Vertical movements** are due to radial forces acting at a marked angle to the surface. They include large-scale uplifts or depressions on a continental scale known as epeirogenic movements. Isostatic movements are similarly included (although one could argue that the final isostatic movement of a mountain range was an orogenic process).

The lateral movements (folding, faulting, and rifting) described earlier were all the result of tension or compression acting laterally at the surface of the crust. The cause of this tension or compression, however, is usually a vertical movement. For example, a section of the crust will bulge or up-arch due to a rising plume of magma or downwarp as the hot spot cools, or as a neighbouring section rises. Isostatic depression and recovery due to glaciation also cause vertical movements just as isostatic uplift due to prolonged erosion of mountainous areas may be apparent.

Finally, large-scale folding, such as in the still-rising Himalayas, is the result of lateral plate movement, yet the effect appears to be a vertical movement as the crust thickens and rises.

Earthquakes

An earthquake is a seismic shockwave, or more commonly a series of shockwaves causing shaking, oscillation, and even fracture of the crustal surface. Stresses build up along plate boundary and fault zones due to friction. When the strength of the rocks is suddenly overcome, the rocks fracture and shockwaves will radiate from the **hypocentre (focus)** which is the origin of the earthquake, usually deep underground. The **epicentre**, on the surface directly above, will receive the greatest ground shaking, usually for less than a minute, and then weeks of aftershocks as the crust settles.

Study of the seismic waves has revealed much, not only about earthquakes, but about the internal structure of the earth.

P (Primary) waves reach the surface first. They travel through both mantle and core and may be measured on the opposite side of the planet. The P waves 'push' – rather like snooker balls in a line.

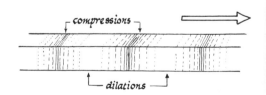

S (Shear) waves reach the surface second. They will travel through the mantle, but not the core and so cannot be measured opposite the event. The S waves 'shake' like a skipping rope.

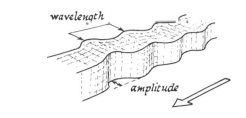

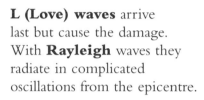

L (Love) waves arrive last but cause the damage. With **Rayleigh** waves they radiate in complicated oscillations from the epicentre.

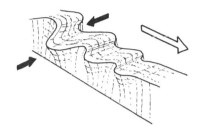

Earthquake vibrations are measured by sensitive instruments called **seismographs**.

The **Richter scale** measures the total power or energy of an earthquake. The scale is logarithmic, extending, theoretically, from zero to infinity. In practice, however, 1 to 9 is measured, each unit being ten times the power of the number preceding it. Earthquakes exceeding 6 are described as 'important', 7 'major', and 8 'serious'.

The **Mercalli scale** relies upon observation of damage to measure the intensity of the earthquake. The scale ranges in single units from I (imperceptible - measured only by instruments), through IV (moderate - likely to rattle windows) and VIII (destructive - chimneys fall), to XII (major catastrophe - complete destruction).

Earthquakes may trigger mass movements such as landslides and allow release of magma through surface fissures. **Tsunamis** (misleadingly popularised as 'tidal waves') may be generated by sea floor earthquakes and cause catastrophic coastal flooding. There is more to earthquakes, therefore, than simply collapsed buildings and damaged infrastructure. Indeed, as with so many natural hazards, the immediate human consequences may be exacerbated by protracted disruption to essential services such as fire control, shelter, water, sewerage, and medical provision - not least maintenance of civil order. As a result, appreciation of causes and consequences with a view to heeding the lessons learned, so enabling **contingency planning** for disaster prevention and mitigation, remains the analyst's principal objective when studying earthquakes or any other natural hazard.

As for predicting them, various preliminary events have been studied such as microquakes before the main tremor, dilation (bulging) of the crust, changes in the electrical resistance and magnetic fields of local rocks, soil argon gas increases, and even curious behaviour in some animals. However, not all of these occur all of the time - nor in the same place - nor even with a regular time lapse before the main event! Prediction is problematic, therefore, to say the least. Likewise earthquake control might appear fanciful, although experimentation in fault lubrication using injections of petroleum drilling 'mud' is proving rewarding in California.

Vulcanicity

Vulcanicity refers to all the processes whereby **magma** (the molten material of the mantle) is forced into the crust. Zones of weakness are exploited by the magma, at high temperatures and pressures, which may even erupt onto the surface as **lava**. Volcanic activity is, therefore, classified as **intrusive** or **extrusive** depending on whether it breaks the surface or not. It is also closely related to tectonic plate margins and earthquake zones.

Intrusive volcanic features cool, crystallise, and solidify into igneous rocks at depth below the surface. Slow cooling results in large crystals forming, as characteristic of, for example, granite and dolerite, which may only become part of the landscape if subsequent erosion removes overlying rocks.

◆ **Dykes** form where magma solidifies in a vertical fissure and may be associated with 'swarms' such as on Arran. If the dyke material is, as normal, more resistant than the surrounding (country) rock, it would be left as a prominent wall-like feature on an eroded landscape such as the Cleveland Dyke. Less resistant dyke material leaves ditch-like features, following erosion, such as the Whin Dyke on Arran, where the country rock is very hard granite.

a few metres to
many kilometres

a few centimetres to
many metres

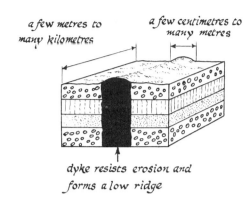

dyke resists erosion and
forms a low ridge

depression

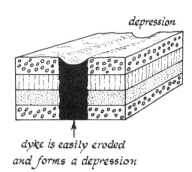

dyke is easily eroded
and forms a depression

◆ **Sills** are solidified magma in horizontal or inclined sheets, for example, the Great Whin Sill in Northumberland which forms steep north-facing cliffs of coarse-grained dolerite where it breaks surface. Indeed, Hadrian's Wall harnesses the natural defensive advantage of this feature in places.

low escarpment

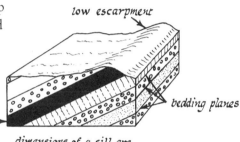

bedding planes

sill

dimensions of a sill are
similar to those of a dyke

◆ **Laccoliths** occur where viscous magma forces overlying strata into a dome, such as found in the Henry Mountains of Utah, USA.

◆ **Batholiths** are more important, however, and notably massive in size and depth. Indeed, often dykes, sills, and laccoliths feed off what becomes the normally granitic batholith before it solidifies. The Goat Fell complex on Arran is an excellent example.

laccolith

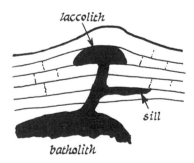

sill

batholith

NB: In all the above features, metamorphism of the country rock adjacent to the intrusions is to be expected.

◆ **Necks** and **plugs** are solidified lavas which have cooled within the volcano's vent. Once the surrounding softer ashes of the cone are eroded away, they may be left as prominent protrusions, such as Edinburgh's Castle Rock or the spectacular defensive site at Le Puy in the Auvergne.

Extrusive volcanic features cool, crystallise, and solidify from surface lavas. Cooling is far more rapid, hence the resulting igneous rocks tend to be finer grained, with small crystals, such as basalt. Both the type of eruption and the nature of the ejected material will determine the resultant feature.

Ejected material may be gaseous, solid, or liquid. Gaseous emissions are dominated by steam. Solids include blocks of material previously 'plugging' the vent, along with dust and ash. Liquids include lava bombs (known as tephra or pyroclasts) which solidify in mid-air and lava (either acid or basic) flowing from both vents and fissures.

◆ **Acid (andesitic) lava cones**: Andesitic lava is viscous, with much silica, and consequently slow flowing. These lavas are associated with subduction where oceanic crust is being destroyed along convergent tectonic plate margins. Volcanic cones tend to be domed with convex sides. Due to the lava's gassy potential, yet high viscosity, it is prone to violent eruptions, especially if so thick that it is unable to flow. Explosive outbursts of incandescent gases, ash, and pumice, known as **nuée ardentes**, are not uncommon, such as witnessed repeatedly on Mount Pelée on the Caribbean island of Martinique.

◆ **Basic (basaltic) lava cones**: Basaltic lava is more fluid and fast flowing. Dominated by iron and manganese, it is associated with mid-ocean ridges and other circumstances where magma has direct access to the surface. It forms flatter cones and shields such as those forming the Hawaiian Islands (a rare example of a **hot spot** distant from any tectonic plate margin where the crust is so thin that magma has 'scorched' a route through).

◆ **Composite cones** are formed from repeated eruptions emitting successive layers of ash and/or lava. The classic conical shape of Japan's Mount Fuji is relatively rare compared to the many irregular composite volcanoes, such as Mount Etna in Sicily. These are constructed from alternating eruptions of tephra and lava and may be covered by numerous secondary (parasitic) cones. The shape of the volcanic cone reflects the nature and composition of the eruptions – hence both form and eruptive behaviour may be used to categorise them.

Section

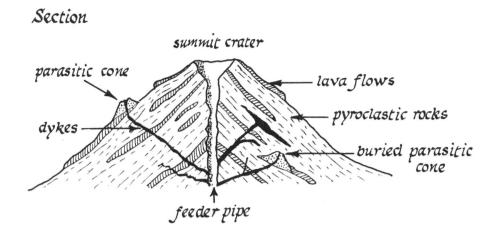

summit crater

parasitic cone

lava flows

pyroclastic rocks

dykes

buried parasitic cone

feeder pipe

♦ **Craters** are the exit vents, usually blocked by 'plugs' of solidified lava in dormant volcanoes, but occasionally liquid as in Hawaiian ones.

♦ **Calderas** are vast pit craters up to many kilometres in diameter. They result from violent eruptions blowing the volcano's summit off, clearing the magma chamber and so allowing the sides to collapse inwards.

♦ **Crater** and **caldera lakes** in dormant and extinct volcanoes are self-explanatory. However, the former may contain islands associated with subsequent eruptions – such as Wizard Island, a cinder cone in Crater Lake, Oregon, USA.

Volcanic eruptions may be from vents or fissures. Vent features include localised cones, craters, calderas, and crater/caldera lakes. Fissure features tend to be more widespread.

♦ **Fissure** features include **lava plateaux** whereby extrusions have filled hollows in the existing landscape. Blocky (*aa*) or smooth ropy (*pahoehoe*) lava may be involved. Indeed, extrusions beneath the sea, originally filmed in Hawaii, showed 'pillow structures' forming like globules squeezed from a tube of paint! Spectacular horizontal jointing of the resulting basalt can be seen on the Giant's Causeway in County Antrim, Northern Ireland.

Minor volcanic features include **fumaroles** which are usually low pressure outlets of gas and steam (sometimes known as **solfatara** if the gas is sulphurous and leaves sulphur deposits on surrounding rocks). **Geysers** are more spectacular, for the pressure builds up and eventually ejects a jet of steam, often with measured regularity, as in Old Faithful in the USA's Yellowstone National Park. **Hot springs**, such as in Iceland, are not under pressure and the resulting water flow may contain minerals dissolved from the hot rocks essential to the operation of each of these minor volcanic features.

Finally, vulcanicity may be viewed from different perspectives - both positive and negative. **Positive** benefits include the valuable mineral deposits, including gold, silver, and diamonds, along with the more mundane such as sulphur, fertile soils, and tourist potential exploited throughout the world. **Negative** aspects tend to be alarming to say the least - scorching lavas burying and burning, ashes suffocating and polluting, dust emissions 'seeding' torrential rainstorms with resulting wet ash and mud **lahars**, and not least flooding following the melting of snow and diversion of natural drainage patterns by lava flows. However, as indicated in the discussion on earthquakes, only non-fatalistic, objective study of volcanic events is likely to further the research necessary if disaster prevention, such as illustrated by the successful diversion of Mount Etna's lava to save Zafferana in 1992, is to be more widely perceived as possible.

The rocks of the crust

There are three main categories - igneous, sedimentary, and meta-morphic.

Igneous rocks are solidified from molten magma. The rock that results will depend both on its chemical composition and the circumstances of cooling. For example, more than 65 per cent silica will give an acid rock such as granite - light in colour and weight. Less than 55 per cent silica will give a basic rock such as dolerite or basalt - characteristically dark and heavy. Those which cool slowly at depth (plutonic rocks) grow large crystals, such as in granite. Rapid surface cooling (volcanic rocks), by contrast, only allows limited crystal growth resulting in fine, dense textures, such as in basalt.

Sedimentary rocks are normally formed from pieces of igneous or organic remains, laid down in layers called **strata**. Sedimentation often takes place under water on ocean or lake beds, later to be uplifted. Some takes place on land. Between the strata are **bedding planes** which are unique to sedimentary rocks and mark any changes in the conditions of sedimentation. It is of particular note that sedimentary rocks are the only category to contain fossils. When these rocks are uplifted the strata may be folded or tilted from the horizontal - a difference known as the angle of dip.

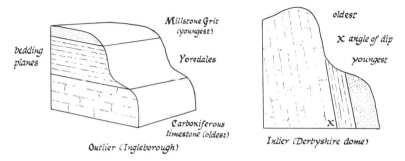

Outlier (Ingleborough)   Inlier (Derbyshire dome)

NB: An **outlier** is a hill or knoll younger than the surrounding rocks. An **inlier,** by contrast, is older than its surroundings.

Sedimentary rocks may be mechanically, organically, or chemically formed and are classified as such.

SEDIMENTARY ROCK CLASSIFICATION

	Classification	**Material**	**Rock type**
Mechanically formed of cemented or compacted fragments	Argillaceous	Mud	Clay/Shale
	Arenaceous	Sand	Sandstone
		Grit	Gritstone
	Rudaceous	Rounded fragments	Conglomerate
		Angular fragments	Breccia
Organically formed from plant or animal remains	Carbonaceous	Peat	Lignite/Coal/ Anthracite/Graphite
	Calcareous	Shells/ Skeletons	Limestone/ Chalk
		Coral	Limestone
Chemically formed		Calcium carbonate on moss	Tufa
		Geyser/Hot spring deposits	Sinter/Travertine

Metamorphic rocks are any others which may have been altered due to extreme heat and/or pressure. Metamorphism may occur by direct contact at a small-scale, such as shale turned to slate along the North Craven fault. Conversely, very large-scale (regional) metamorphism is associated with the edges of batholiths. Common examples of metamorphism include chalk and limestone turned to marble, sandstone to quartzite, clay and shale to slate, and granite to gneiss.

Joints and unconformities

Joints may occur in all three rock categories. They are **cracks**, usually found at an angle to the strata, caused by tension. They are associated with molten igneous and metamorphic rocks solidifying and shrinking. Likewise, sedimentary rocks dry and shrink. Finally, earth movements may cause the tension responsible.

Unconformities occur when younger rocks are deposited on top of an ancient erosion surface, such as illustrated by Carboniferous limestone overlying Silurian shale in the Yorkshire Dales. In this case the Devonian rocks which originally covered the shale were eroded away before the Carboniferous age (see earlier diagram of the Mid-Craven fault).

2
WEATHERING AND SLOPES

Geomorphology is the study of the earth's landforms. It is concerned with analysis and explanation of their shape, the erosional and depositional processes at work on them, and their evolution through time. This timescale may be diurnal (daily), seasonal, annual, or longer. Indeed, geological timescales are incomprehensible, especially when one considers the planet as 4600 million years old with a fossil record covering only the last 600 million years. Over one million years of human evolution can be argued, but only in the last ten thousand years have we influenced geomorphological processes - mere 'seconds' in geological time, but of ever increasing influence nonetheless.

When considering the shaping of landforms the distinction between internal and external processes is important to appreciate. Endogenetic processes (discussed in *Earth Structure*) stem from tectonic forces beneath the earth's surface. **Exogenetic (external) processes**, however, relate to surface weathering and the erosive action of water, wind, ice, and the sea. These processes effectively sculpt the endogenous features.

SELECTED DEFINITIONS

Denudation is a general term for the wearing down of the earth's surface. Weathering, erosion, and transport are involved - collectively breaking rock down and taking away the debris.

Weathering is the decay and disintegration of rocks at or near the earth's surface, *in situ*.

Erosion refers to the wearing away of the land surface and removal of debris by river and sea water, ice, and wind.

Deposition refers to the accumulation of the sediments and fragments formed by weathering and erosion. These sediments may be deposited by water, ice, and wind or be evaporated residues, such as salt, or even chemical precipitates, such as tufa and sinter.

Weathering (mechanical and chemical)

Weathering is the necessary initial phase of denudation, providing the products which form the **regolith** (a collective name for the soil and rock waste covering the earth's surface). This material may either be left *in situ* or removed by gravity and agents of erosion such as water, wind, ice, or the sea.

Mechanical (physical) weathering is the fracture, and so breakdown, of rocks into fragments. It may be caused by frost shattering, pressure release, salt crystallisation, thermal expansion, or even biologically.

1. **Frost shattering** is the most important form of mechanical weathering, occurring where the temperature of exposed rocks oscillates around freezing point, such as in upland areas. For example, water entering any joints in daytime may freeze at night, and so expand by c. 10 per cent. Regular **freeze-thaw** cycles would progressively widen the joints until fracture occurred. The disintegration of these shattered blocks builds the **screes** or **talus** beneath steep slopes, and **blockfields** or **felsenmeer** on gentle relief, found throughout upland Britain.

2. **Pressure release (dilatation)** occurs when rocks such as granite, which were developed under considerable pressure, are exposed to the atmosphere by the erosion of overlying rocks. The burden of weight is, therefore, removed – so allowing any stress to be released in the form of expansion. Consequently, cracks may develop parallel to the surface causing **sheeting**, as seen on granite tors on Arran.

3. **Salt crystallisation** cannot produce fractures, only enlarge existing ones by saline water in joints or pore spaces evaporating to leave salt crystals forming. Capillarity in deserts is a common cause and coastal zones are vulnerable too. At the scale of individual grains breaking off (**granular disintegration**), the stonework of buildings can be progressively denuded. For example, York Minster is built of magnesium limestone which reacts with the sulphur dioxide in acid rain to grow magnesium sulphate crystals.

4. **Thermal expansion** on heating, and contraction on cooling, have long been quoted to explain the diurnal stresses breaking up the surface of desert rocks in what was called 'onion skin weathering'. However, recent

experimentation has suggested that this **exfoliation** is more likely to be due to hydration and crystallisation processes. Also, at a granular scale, thermal expansion and contraction at different rates, according to the variety of minerals involved, was assumed to be responsible for sand formation in deserts. However, experimentation now suggests that this process is ineffectual without water. Continuing research in physical geography is, therefore, downgrading the role of thermal expansion as a mechanical weathering process.

5. **Biological (mechanical) action** is also of less significance than once thought, especially relative to its chemical aspects, in that roots will follow the line of least resistance rather than exploit joints at will. However, root expansion will widen joints, as will the leverage exerted by trees swaying in strong winds. Burrowing animals will excavate material, such as partially weathered rocks.

Chemical weathering involves decomposition of rocks - literally changing the minerals. It is most effective in areas with high humidity because rain water is (naturally) a weak carbonic acid and water is needed as a chemical reactant and transporter. Also, high temperatures are required because chemical reactions will be faster. Indeed, there is a two and a half times increase in reaction speed for every 10°C rise in temperature. Note, however, that chemical weathering often operates in conjunction with mechanical weathering and so hydrolysis, oxidation, solution, carbonation, hydration, and organic (biological) processes may be difficult to differentiate from physical ones.

1. **Hydrolysis** is probably most significant because it decomposes minerals, such as feldspar in granite, to form clay. The process is the same as cation exchange in soil development - H_2 in the water replacing cations in the mineral (see *Soils*). Hydrolysis is speeded up by the increasing prevalence of acid rain in urban industrial societies.

2. **Oxidation** is when iron components in a ferrous state join with oxygen into the (rusted) ferric state. **Reduction** - the reverse process - may operate in waterlogged areas. Both processes are evident by distinctive localised colouring of the minerals involved - red in the former, blue/grey in the latter.

3. **Solution** and **carbonation** in humid areas accounts for the direct removal of more material than any other process! Solution is the dissolving of minerals by pure water. However, rain, as mentioned earlier, is weak carbonic acid and carbonation is, therefore, far more significant and particularly effective in limestone areas. The rock is dissolved and removed in solution as calcium bicarbonate. The limestone may then be reprecipitated as, for example, cavern features such as stalactites and so on, or as tufa on moss.

4. **Hydration** is a physio-chemical process whereby water is absorbed by certain minerals (a chemical process) causing them to expand and exert (physical) pressure. Alternating hydration and dehydration, according to the weather, may be particularly damaging to the foundations of buildings.

5. **Organic (biological) processes** relate to the acids produced by bacteria, fungi, algae, and guano. Humic acid from decomposing vegetation is released by a process known as **chelation**. These processes are of particular note by their collective effect of increasing the strength of chemical reactions.

Processes of chemical weathering

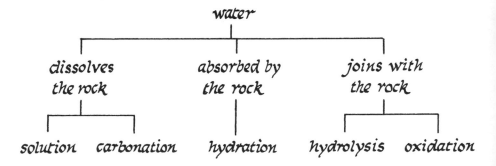

FACTORS AFFECTING WEATHERING

1. **Mineral composition:** Acid rocks, containing more resistant quartz, are less susceptible to weathering than basic ones.

2. **Texture:** Coarse-grained rocks are more susceptible to weathering because the decomposition of one mineral weakens the whole piece more than it would in a fine-grained sample.

3. **Climate:** Cold climates inhibit chemical weathering whereas in humid temperate ones, it will be far more important than mechanical processes. In hot deserts we now know exfoliation to involve hydration, and oxidation can lead to 'painted deserts'. Humid tropical conditions are ideal for excessive chemical weathering - rotting rocks to depths of 60 m!

4. **Time:** The degree of weathering increases as the duration of exposure is extended. However, **negative feedback** may operate by a thick mantle of debris protecting the rock from further chemical weathering, unless it is removed.

5. **Lithology (the physical characteristics of rocks):** Granite, for example, is vulnerable due to its jointing, yet basalt weathers quicker because it is basic, and so less acid. The resistance to chemical weathering of sedimentary rocks depends on their permeability and the nature of the cementing material.

6. **Nearness to sea:** Although the pH of the sea is neutral, salt crystals and the chemical action of sodium are of significance.

7. **Human activities:** Acid rain, aggravated by CO_2, SO_2, and NO_2 from power stations, vehicles, and industries will accelerate chemical weathering, particularly in urban areas.

Slope processes

There are few places where the land surface is truly horizontal. Consequently, landscapes may be thought of as made up of slopes. These are dynamic - always changing in response to the following processes:

Mass movement is the general term for the movement of material, such as soil, loose stones, and rocks, by gravity. The presence of water is important in order to act as a lubricant.

Gravity has a dual effect in both sticking particles to a slope, yet sliding them down too. As a consequence, slope angles are often said to result from a *'battle between stick and slide'*. This relates to the basic forces acting on any particle on a slope. Providing the 'stick' component is greater than the 'slide' one, then there will be no movement. Should, however, the angle be steep enough for the 'slide' component to dominate, then movement will occur.

Simple trigonometry allows the 'stick' and 'slide' force components to be calculated, providing the particle mass and angle of slope are known.

The **angle of friction** is the slope angle at the point where frictional forces exactly balance the slide forces trying to pull the particle downslope.

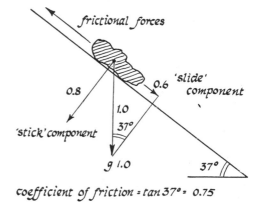

The **coefficient of friction** is the tangent of this angle – the ratio between the 'stick' and 'slide' components.

coefficient of friction = tan 37° = 0.75

After A. Clowes and P. Comfort (1982)

However, water filling pores between the particles will force them apart. This **pore pressure** will affect the cohesion of the material, especially at low angles.

The angle at which debris of any given size is naturally stable is known as the **angle of repose**. The larger the particle size, the steeper this angle and so the greater its coefficient of friction. Fine materials cannot rest on a slope greater than 25 degrees, whereas a rock scree's angle of repose may be up to 60 degrees.

However, uniformity of particle size
is less likely than the mixture found
on many scree slopes. This is further
complicated by numerous small particles lodged between bigger ones on
the steeper sections - and larger pieces
on the gentle slope, having rolled down.

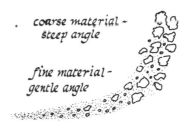

Mass movements may be slow or rapid, periodic or continuous. Four main
processes are distinguished - creep, flow, slide (rotational and non-
rotational), and fall.

1. **Creep**, whether **soil creep** or **rock creep**, is a very slow, but usually
continuous process. Consequently, it is imperceptible and so recognised
only by its effects. It occurs on most slopes steeper than 5 degrees. Whilst
rates between 5 and 10 cm per annum may be enough to tear the top cover
into rolls and step-like **terracettes**, especially on angles exceeding 30
degrees, normally creep would not exceed 1cm a year.

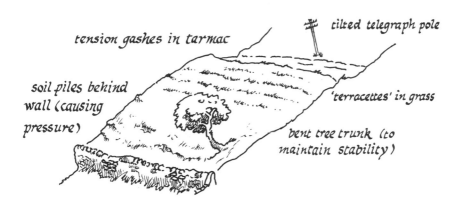

Creep results from repeated expansion and contraction of the material. For
example, freeze-thaw cycles will **heave** particles up on freezing and allow
them to fall further down slope on melting. Alternate hydration and
dehydration will cause expansion and contraction of particles. However,
because the expansion widens the sloping rest bed, the particle can slip
lower into the enlarged depression
on contraction. Thermal expansion
and contraction in the absence of
water would have the same effect.

expansion and contraction

2. **Flow** is more rapid than creep, but sporadic rather than continuous. Most characteristic is the velocity decrease from surface to base. Earthflows, mudflows, and solifluction are differentiated.

◆ **Earthflows** contain less than 20 per cent water. Slower and more viscous than mudflows, they often follow distinct **flow tracks** with bulging **lobes** at the toe marking the final resting point. The 1966 Aberfan disaster (which, arguably, might better be described as a mudflow given the rapidity of the event) is a much-quoted example, the lessons from which have enabled the mining industry to stabilise other colliery spoil tips by drainage and maintenance of height and slope.

◆ **Mudflows** contain 20-50 per cent water and are, therefore, faster and capable of an enormous carrying capacity. The Armero tragedy following the 1985 eruption of Nevado del Ruiz in Colombia illustrates, again, their potential for destruction. This flow, for example, was estimated to reach speeds of 80 km per hour!

◆ **Solifluction** (literally meaning 'soil flow') is a term best used for the periglacial process of slow downhill movement on top of frozen subsoil. Rounded, tongue-like solifluction lobes are common summer events in permafrost regions of limited vegetation, such as the tundra. Although still found in the Cairngorms, most solifluction **head** deposits in Britain are relicts from the Pleistocene Ice Age, 10 000 years BP.

3. **Slides** are generally rapid and sporadic. The movement is known as **constant flow** because the entire mass moves as one (unlike the decreasing velocity from surface to base noted in flows). Rotational slides should be differentiated from non-rotational ones.

◆ **Rotational slides** are sometimes called **slips** or **slumps.** They are characterised by a back tilted slope and may have an earthflow emerging from the toe. Coastal erosion is a common cause with waves undercutting the base of clay saturated by heavy rain, so triggering the rotational slide. Complex slips within slips may be apparent too, especially on vulnerable dipping strata.

◆ **Non-rotational slides** are described exactly in their name - identical to the above, but without rotation. Italy's Vaiont Dam disaster in 1963 was a horrifying example of a non-rotational rock slide inundating a reservoir with fearful consequences.

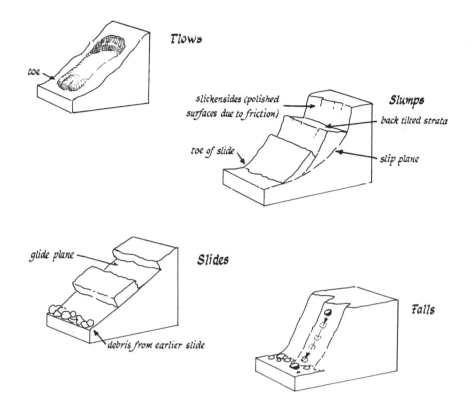

4. **Falls** are associated with near vertical slopes (70–90 degrees) where there is no 'stick' component. Clearly, only very hard, resistant rocks could sustain such angles on coastal cliffs, quarries, and fault scarps. Slope undercutting and frost shattering are particularly frequent 'triggers'.

Removal by running water on slopes

◆ **Rainsplash** (raindrop impact) both dislodges particles and undermines slopes.

◆ **Sheetflow** (an overland flow on a saturated surface) may erode unvegetated slopes.

◆ **Rills** and **gullies** (whereby flow is concentrated into channels) will, likewise, erode saturated, unvegetated slopes.

Slope form

As stated earlier, all landforms are made up of slopes, each of which can be thought of as a system of **input** (material gained by weathering), **output** (material lost by transport), and **storage** (whatever is left). The concept of **dynamic equilibrium**, whereby the processes operating on any slope will move towards a balance is, arguably, the key to understanding slope form. This is only difficult in limestone areas where the landscape is lowered without significant transport being visible because the transporting medium is solution.

Different elements of a slope are identified, such as **facets** and **breaks of slope**. These may be plotted on morphological maps or identified on slope profiles.

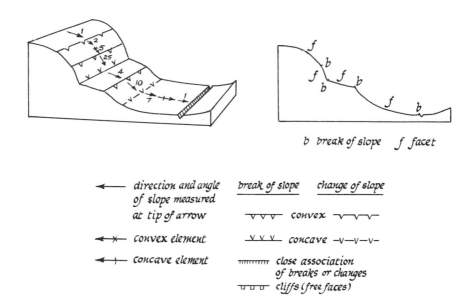

b break of slope   f facet

←— *direction and angle of slope measured at tip of arrow*	*break of slope change of slope*
←×— *convex element*	⌄⌄⌄ *convex* ⌄⌄⌄
←╁— *concave element*	∨∨∨ *concave* –∨–∨–∨–
	⊓⊓⊓⊓ *close association of breaks or changes*
	⊓⊔⊓ *cliffs (free faces)*

In humid temperate climates, such as in Britain, most slopes display an upper convexity and lower concavity. The upper slope is usually a convex **slope of denudation** – with increasing steepness with distance from the summit necessary to remove all weathered material.

Assuming the removal of a layer of equal thickness from the whole summit, then progressively more material is added and must be moved downslope as the distance from the summit increases.

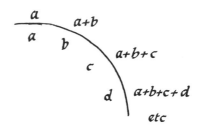

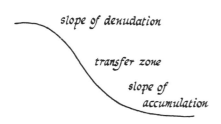

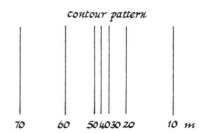

The mid-sections of the slope should be regarded as the **transfer zone**. For example, it could be a cliff where free fall was apparent with a straight 'debris' slope beneath actively transporting materials by creep and slide. Finally, the concave lower slope, becoming less steep towards the bottom, is the **slope of accumulation**.

Particularly important to note is whether or not there is an active channel at the base of the slope, for this affects the rate of debris removal.

Clearly the next stage is to consider the third dimension. In plan, **gathering** (funnel-shaped) **slopes** are concave, whereas **spreading slopes** (opening out at the base) are convex. However, profiles must be regarded. Convexity reflects the creep slopes of upper profiles whereas concavity typifies lower slopes.

Gathering slopes   Spreading slopes

concave plan
convex profile

concave plan
concave profile

convex plan
convex profile

convex plan
concave profile

FACTORS INFLUENCING SLOPE FORM

1. **Climate** will directly affect soil expansion and contraction through freeze-thaw and wet-dry periods. Heavy precipitation in the form of rain or snow may be literally that - heavy! Even indirectly, climate influences vegetation, which (as mentioned below) can protect slopes.

2. **Lithology** affects the strength of slopes. Clay, for example, is relatively weak, and stable only up to 10 degrees. However, most other rocks are stable up to 35 degrees. Permeable lithologies would prevent saturation of the topsoil and so improve slope stability. Even solubility is relevant given the smooth, rounded profiles in, for example, chalk compared to angular limestone.

3. **Structure** is particularly relevant when resistant rocks such as limestone alternate with softer ones like clay, so increasing vulnerability to movement.

4. **Soils** prone to saturation, and so pore water pressure forces, are less stable, as are thin soils with limited vegetation and so poor root binding.

5. **Vegetation** not only binds soil together but intercepts precipitation, so reducing surface runoff.

6. **Human activities** influence slopes to varying degrees. Deforestation, overgrazing, and failure to contour plough, poorly designed cuttings and embankments, quarrying and tipping - not least building weights and traffic vibrations - all have relevance to slope stability.

Slope evolution

Slope evolution is a complex and controversial topic because the processes are both very slow and varied. Indeed, the combination of processes, let alone environmental variations influencing each slope, will differ widely across the world, so prohibiting generalisations. **Slope evolution theories**, consequently, are highly speculative and should be regarded with caution.

It is of note that the earliest theories were all **inductive** (created to fit the observed evidence). Later theories, however, were **deductive** (starting with a hypothesis which was then tested in the field).

SLOPE DECLINE THEORY (W.M. Davis, 1899)

This postulates slope recession with a lowering of the entire surface. **Stage 1** sees falls and slumps on the irregular initial slope. **Stage 2** is the 'graded slope' of less than 35 degrees with regolith cover throughout. **Stages 3** and **4** see progressive softening of the upper convexity and lower concavity, the former of which will eventually extend to flatten the whole slope out.

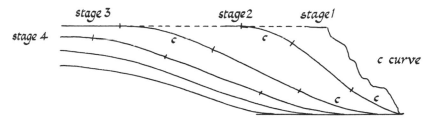

In effect, the upper surface (slope of denudation) is progressively lowered by creep, surface wash, and solution. The middle section (transfer zone) collects this and its own debris, whilst the lower section (slope of accumulation) collects it all. This protects it from further weathering and lowering - providing there is no active channel there to remove the material.

The theory is inductive - based upon observations in humid temperate climates and on the relationship between weathering and transport. It takes no account, however, of lithology, structure, or climatic differences.

SLOPE REPLACEMENT THEORY (W. Penck, 1924)

This envisages originally steep slopes being replaced by lower angle ones extending from the base at constant angles. **Stage 1** sees the free face cliff slope (A) slowly buried by an accumulating scree (B). **Stage 2** sees this scree replaced by a lower angle slope of small particles washed from it (C). **Stage 3** sees the whole process continuing until the scree (B) replaces the cliff entirely, with the gentler wash slope (C) spread below.

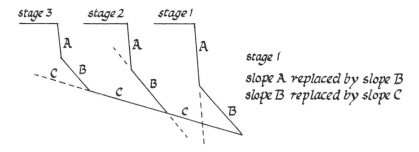

PARALLEL RETREAT THEORY (L.C. King, 1948, 1957)

Here the slope facets remain in their original proportions as it retreats. The profile, therefore, remains unchanged, with only the base extending. This would be a pediment in an arid environment or the slope of accumulation in a humid temperate. one. At the coast it would be the inter-tidal wave cut platform. Structure seems to be important in the maintenance of parallel retreat. Horizontal bedding, for example, may be retained until the protective caprock is removed.

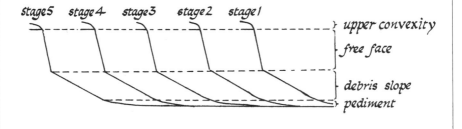

3
ROCKS AND RELIEF

The inquisitive, but untrained, might view a landscape and assume that type, form, and strength of rocks determines the relief. Certainly, the fascination of early physical geographers with granite and limestone landforms led to rock type as the factor dictating relief being stressed. However, rock types and structure are just part of a plethora of factors determining relief. Whilst significant, therefore, they must not always be assumed to be of paramount importance. Tectonic forces, weathering and erosion, in conjunction with rock characteristics, operating under different environmental circumstances determine different landscapes. Indeed, rock types may, occasionally, be of limited relevance – as in the case of arid landscapes.

Clearly, however, different rock types may be associated with distinctive scenery. **Lithology** is frequently, therefore, relevant to our understanding of landscape because it determines vulnerability to mechanical and chemical weathering processes and to agents of erosion such as ice, water, and wind. Brief examinations of granite, limestone, and arid characteristics should illustrate, however, just how variable the relationship between rocks and relief can be.

Granite characteristics

Granite is an intrusive igneous rock originating in batholiths. It may be hard but does not necessarily control landscape because different environmental conditions result in different geomorphological processes.

The **tor**, as found on many Dartmoor summits, is granite's most characteristic landform. Two main theories have been proposed in order to explain these features. D.L. Linton emphasised sub-surface chemical weathering during warmer conditions before the last Ice Age, concentrated where jointing was closest. J. Palmer and R.A. Neilson, however, suggested a major role for sub-aerial frost shattering under colder periglacial conditions. Whatever the original weathering process, both explanations suggested removal of most of the core stones, along with the protective overlying rocks, by solifluction processes during periglacial conditions at the end of the last Ice Age. The tors left upstanding, therefore, are relict features and exposed to further chemical, but predominantly mechanical weathering, in the form of frost shattering.

Limestone characteristics

The relationship between limestone and resulting landscape is very close. Indeed, rock type is the dominant controlling factor with the carbonation process dictating **karst** scenery in areas of widespread and thick limestone, such as in the former Yugoslavia. Regions of limestone in England, such as the Peak District and Yorkshire Dales, by contrast, might best be described as displaying **karstic** features.

Carbonation involves the dissolving of limestone by weak carbonic acid. A weak carbon dioxide (CO_2) solution in water, such as soil water given the abundance of CO_2 in soils will, therefore, attack the limestone and dissolve it.

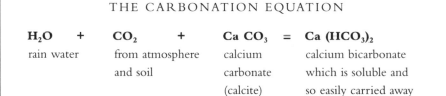

THE CARBONATION EQUATION

H_2O	+	CO_2	+	$Ca\ CO_3$	=	$Ca\ (HCO_3)_2$
rain water		from atmosphere		calcium		calcium bicarbonate
		and soil		carbonate		which is soluble and
				(calcite)		so easily carried away

NB: The process is reversible with reprecipitation of **$CaCO_3$** as stalactites, stalagmites, and tufa, for example.

The more lime the water absorbs, the less 'aggressive' it becomes – hence less solution deeper underground. But the amount of CO_2 in the water is the key to carbonation. Since CO_2 dissolves more readily at lower temperatures, erosion rates vary considerably. For example, 5 mm per century in the Yorkshire Dales today compared to 20 mm under periglacial conditions.

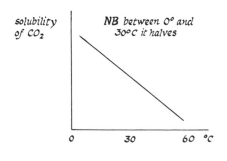

The effect of this process depends on the lithology and structure of the three main types of limestone – Carboniferous, Jurassic, and Cretaceous (chalk). Each of these is associated with its own distinctive landscape, the similarities and differences of which may be summarised by considering:

◆ surface drainage,
◆ surface solution features,
◆ collapse features,
◆ subterranean features,
◆ slope composition.

A comparison of Carboniferous limestone and chalk, for example, illustrates this sequence well. **Jurassic (oolitic) limestone**, formed 180 million years BP, however, is too variable in character to allow productive generalisation.

1. **Carboniferous (mountain) limestone** was formed 350 million years BP. Rising to over 300 m in the UK, it is the remains of calcareous organisms including many impurities. Massive, hard, resistant, and well jointed, it displays good **secondary permeability** (whereby erosion is selectively concentrated along lines of weakness such as joints and, if tilted, bedding planes).

◆ **Surface drainage** is usually intermittent and fragmented and may even be entirely absent. Some rivers may persist, however, because they are at the level of the water table. For example, Gordale Beck near Malham in the Yorkshire Dales flows because of a 'locally perched' water table. All drainage must have started on the surface, as suggested by the existence of dry valleys, but then has gone underground once overlying impervious layers were removed by erosion. Certainly, dry valleys would suggest previously higher water tables – the valleys having been cut in periglacial conditions when the subsoil was frozen and so impervious. The Watlowes above Malham Cove, for example, has upper and lower dry valleys separated by a dry waterfall.

◆ **Surface solution features** may be classified as closed or open. Closed depressions formed by solution of a master joint vary in size from small, shallow **shake holes** (c. 1 m across) to larger, deeper **dolines** (c. 10 m – 1 km across). Open depressions may have started as a doline only to be enlarged and sometimes cleared of debris by the erosive action of runoff. They are **water sinks** such as the cluster of small, glacial debris filled holes found at Malham Sinks, or spectacular **swallow holes** such as the 160 m deep Gaping Ghyll near Ingleborough.

Pavements probably originated beneath regolith only to be exposed later by, for example, glacial erosion such as above Malham Cove. The rocks must be horizontally bedded in order for the upper surface of the bedding plane to form **clints** separated by **grykes** (dissolved by surface solution down joints).

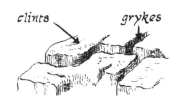

Limestone pavement

◆ **Collapse features** may not always be obvious as such. Certainly, dolines and swallow holes may be enlarged in this way and many steep-sided **gorges** are likely to have been formed by cavern collapse. For example, Cave Dale near Castleton in the Peak District and the head of Goredale, near Malham. Gorges are always controversial, inviting endless speculation as to whether they are simply collapse features, or glacial meltwater channels, or perhaps even the result of waterfall recession (both the latter under periglacial conditions). However, the excessive undercutting at these gorge bases could only have been achieved by rivers flowing at high pressure through enclosed caverns. Whether now dry (Cave Dale) or containing misfit streams (Goredale) they were all probably enlarged by high runoff over frozen, hence impervious, limestone following the last Ice Age - and all are areas of weakness in the sense of their complex joint and fault systems.

FIELD SKETCHING

Field sketching is an invaluable technique whereby one identifies and draws key landscape features as an aid to memory and understanding. First, dominant shapes, such as horizons should be drawn. Secondary features, such as rock outcrops, woodlands, or settlements may then be added. Uncluttered annotation, by labelling landforms, features, and, where relevant, processes is essential. The main benefit of the technique is that it makes one think about landscapes and the physical and human influences determining them.

Field sketch examples: Malham area Yorkshire Dales National Park

Gordale looking NE. (between Janet's foss and Gordale scar)

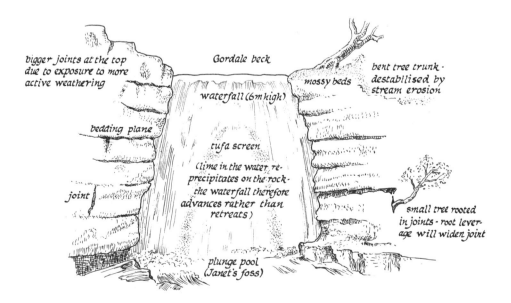

75m high

scars

slope of denudation

active frost shattering

Carboniferous limestone

90°

bent tree trunks suggest slope instability

35-40°   bare rock · limited vegetation cover

slope of transport

30°   scree chutes

terracettes

slope of accumulation

flat bottomed valley · filling in as debris accumulates and cannot be removed by misfit stream (Gordale beck)

Janet's foss

bigger joints at the top due to exposure to more active weathering

Gordale beck

bent tree trunk · destabilised by stream erosion

mossy beds

waterfall (6m high)

bedding plane

tufa screen

(lime in the water re-precipitates on the rock the waterfall therefore advances rather than retreats)

joint

small tree rooted in joints · root lever-age will widen joint

plunge pool (Janet's foss)

Goredale scar (upper Goredale gorge)

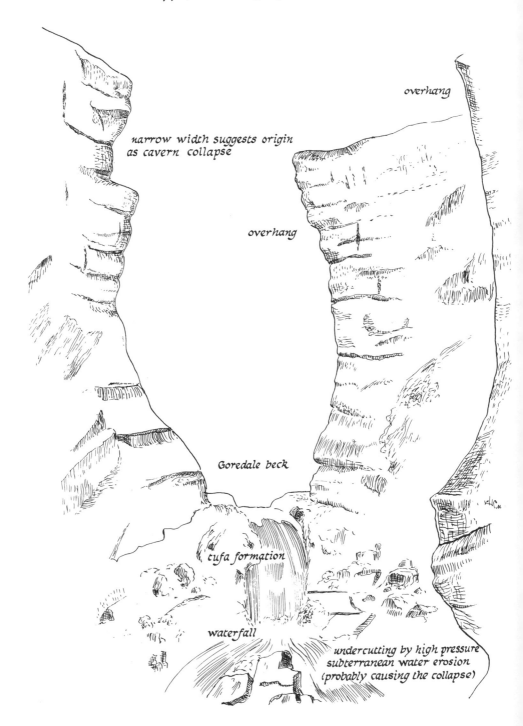

overhang

narrow width suggests origin
as cavern collapse

overhang

Goredale beck

tufa formation

waterfall

undercutting by high pressure
subterranean water erosion
(probably causing the collapse)

◆ **Subterranean features** such as **caves** and **caverns** are, normally, enlarged horizontal fissures, usually following the bedding planes. They were cut either in the **phreatic zone** (below the water table), along the water table, or even at a time when it was higher. These latter examples, such as Treak Cliff Caverns near Castleton, are now left stranded 'high and dry' in the present **vadose zone** (immediately above the water table) and display characteristic depositional features. **Stalactites, stalagmites, pillars**, and **flowstone** represent the carbonation process in reverse as moisture and carbon dioxide return to the cave air, so leaving the calcium carbonate reprecipitating.

◆ **Slope composition** in limestone regions reflects both its strength and tendency to weather into joint controlled blocks. Vertical slopes and steep-sided gorges are, therefore, not uncommon with **scars** (cliff-like features) and **stepped profiles** characteristic where the rocks are horizontally bedded. The multiple slope facets typical of Carboniferous limestone are illustrated particularly well in Gordale Scar near Malham.

2. **Cretaceous limestone (chalk)** was formed 135 million years BP. Rarely over 180 m in the UK, it is the purest form of limestone and made of algal remains. It may be soft (North and South Downs) or hard (Flamborough Head), but is always mechanically weak because of its many irregular cracks, despite being less well jointed. Indeed, it is porous and so displays **primary permeability** (with the water passing through the pore spaces between individual grains). Any fissures, however, allow secondary permeability. This combination of vulnerability to carbonation makes it the most soluble of the three types.

Yorkshire Wolds escarpment (cuesta)

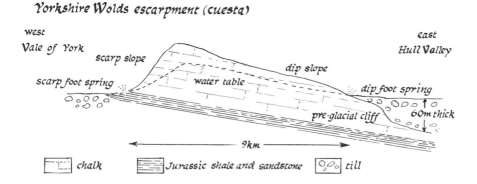

◆ **Surface drainage** is usually absent until the chalk meets impervious rocks where **springs** and **seepage** are likely. For a spring, the water outlet is concentrated along a fissure, otherwise it will just ooze out along a continuous line of seepage. The Yorkshire Wolds demonstrate both scarp foot (Weedley) and dip foot (Beverley Westwood) springs.

Dry valleys, such as Comber Dale, are associated with a variety of formation factors, including meltwater in periglacial streams on frozen bedrock, glacial spillways cutting across existing watersheds, such as at Risby, and scarp recession. **Coombe** depressions on steep slopes are most probably associated with spring sapping (headward erosion) and collapse. **Bournes** are the intermittent streams found flowing down normally dry valleys following heavy rain.

◆ **Surface solution features** rarely form because the primary permeability of chalk allows uniform solution. However, water sinks may occur if runoff is concentrated by the local topography.

◆ **Collapse features** are not found because caves and caverns cannot form in chalk due to its primary permeability and because it is too mechanically weak.

◆ **Subterranean features**, therefore, are absent because fractured chalk would fill them in as soon as they formed.

◆ **Slope composition** is, distinctively, rolling with convex upper slopes and concave lower ones covered in thicker soils than found on limestone. This reflects the softness and solubility of chalk.

NB: Dipping chalk deposits give typical cuesta sequences of dip, scarp, and vale.

Arid characteristics

The introduction to *Rocks and Relief* referred to arid landscapes as illustrative of rock characteristics being of limited relevance in slope development. Yet a third of the world's land surface is subject to the **permanent water deficit** (whereby potential evapotranspiration exceeds precipitation) that best describes aridity. Whilst annual precipitation of less than 250 mm remains the standard definition of desert, variability of this rainfall, as discussed later, lends more clues as to the resulting landforms given the erosive power of heavy storms following protracted drought. Likewise, temperature variability is critical. Annual and, especially, diurnal ranges can be excessive, determined by cloudless skies beneath the descending tropical air of the explanatory **Hadley cell**. This absence of cloud cover ensures no insulation to prevent rapid night–time radiation losses of significant heat energy accumulated under the glaring sun of the day.

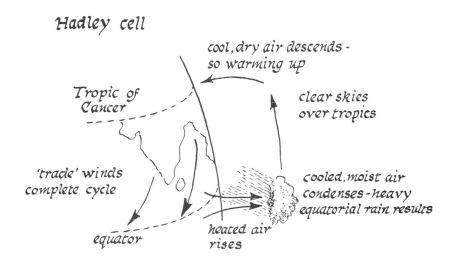

Hadley cell

cool, dry air descends - so warming up

Tropic of Cancer

clear skies over tropics

'trade' winds complete cycle

cooled, moist air condenses - heavy equatorial rain results

equator

heated air rises

But exfoliation of rocks associated with their thermal expansion and contraction is now known to be less significant than physio–chemical weathering processes such as hydration and salt crystallisation, both associated with the occurrence of dew. Similarly, the relevance of wind has been reassessed with satellite remote sensing imagery confirming, for example, **aeolian** (wind) deposition to be central in explaining large-scale dune systems. Clearly, lack of vegetation aids both aeolian erosion and deposition. For example, **deflation hollows**, such as the Qattara depression in the northern Sahara, have lost sand measured in thousands of cubic kilometres! Indeed, rocky (**hamada**) and stony (**reg**) deserts are sometimes explained by deflation processes. Sand in transport is certainly a powerful agent of erosion with **rock pedestals** and elongated, sculptured ridges such as small-scale **zeugens** and larger **yardangs** providing testimony to the abrasive power of creeping, saltating, and suspended sand. But it is aeolian deposition that dictates the most familiar landscapes associated with sandy (**erg**) deserts. For example, linear, narrow **seif** dunes can extend in the prevailing wind direction for over 100 km whilst far rarer crescent-shaped **barchans** migrate individually or in swarms. Meanwhile, more variable forms such as **star** and **echo** dunes reflect complex micro-climatic wind eddies and vortices.

Water in deserts is now understood to be a far more influential landscape variable than previously thought. Limited vegetation interception and ground surfaces, often cemented, of fine particles dictate **sheet wash** and **flash floods** as a normal response to infrequent rain. Steep-sided **inselbergs**, for example, in semi-arid regions project from extensive, gently sloping rock platforms (known as **pediments**). Examined in L.C. King's parallel retreat theory such features could, perhaps, be explained in part by sheet wash and flash flooding. This is suggested because the free faces of inselbergs, prone to both mechanical and chemical weathering, produce screes, but also add to material spread further by flash floods channelled through normally dry ravines called **wadis**. These features are often found dissecting the inselbergs, and when in flash flood, disgorge sedimentary deposits in alluvial fans over the sheet washed pediment below. The occurrence of near flat silt, clay, and salt **playa** (ephemeral lake) deposits at the furthest extent of the pediment, all dried out to leave a crazing of desiccation cracks, lends further evidence to the relevance of water to such landscapes. However, whether the flat-topped **mesas** and (smaller) **buttes** of arid Arizona are similarly explained, or simply represent the degraded remnants of more extensive plateaux, again stimulates enthusiastic debate. It is on this note of speculation that this section should be concluded. As with any landscape, the influencing factors must always be viewed with reference to changes through time. Arid landscapes have not, necessarily, always been dry. How else could the dimensions of wadis and their alluvial deposits be so great, unless far wetter conditions had prevailed in the past?[1] Indeed, **desertification** (the process of desert spread) - one of the world's great contemporary environmental issues - is explained in part by such climatic change. For example, rainfall variability in the Sahel region of Africa has become markedly more acute in recent decades, putting an area twice the size of India at risk to the spreading Saharan sands. But human activities are mostly to blame with accelerating numbers forced to cultivate increasingly marginal land and so reduce essential fallow periods before exhausting soils beyond recovery. Endemic overgrazing and well-intentioned, but often poorly drained, irrigation schemes, causing salinisation, exacerbate the environmental deterioration. This is all made worse by deforestation, for farmland and fuel, which not only removes the soil's protection from sun,

---

[1] Proof of this may be obtained by historical accounts, pollen analysis, oxygen isotope measurements, and so on, used to understand the climatic changes involved in determining glacial landscapes.

wind, and rain, but also contributes to the decline in rainfall by reducing atmospheric moisture because evapotranspiration losses must be less.

Temperature variations, wind, and water, therefore, assume far greater relevance than the rocks themselves in determining the nature of arid landforms.

Rocks and relief, therefore, are undoubtedly related and can never be ignored – but landscapes, even as strongly influenced as limestone, usually reflect far wider issues than simple causal relationships.

4
HYDROLOGY

Hydrology is the study of water as it occurs on, over, and under the earth's surface. This includes precipitation, rivers, lakes, soil moisture, and so on. In spite of the crucial importance of water to life on the planet, hydrology is a relatively young science, with much research of an applied nature emanating from the USA, covering such aspects as conservation of water resources, irrigation, and flood control.

Water occurs in three forms - **liquid** (in rain, lakes, streams, groundwater, and the oceans), **solid** (snow, hail, and ice), and as a **gas** (in water vapour). There is a constant change from one form to another, such as water to ice, and constant movement, such as from groundwater to stream flow.

Oceans contain 97 per cent of all water. Therefore, only 3 per cent is fresh. Of this fresh water, three-quarters is locked up as ice sheets and glaciers with most of the rest as groundwater. Consequently, only a fraction of the world's total water budget is in rivers, lakes, soil moisture, and the atmosphere.

The hydrological cycle

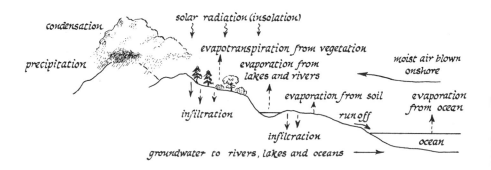

This is clearly an oversimplification in that precipitation may evaporate immediately on reaching the ground (without joining the rest of the cycle). Furthermore, water may join the groundwater and not re-emerge for thousands of years, just as it might be locked up as ice caps in cold climates.

Consequently, whilst this cycle is a useful reference, a **systems approach** allowing measurement and quantitative analysis is required for thorough understanding. Global and local systems are, therefore, differentiated - the former 'closed' in that water is fixed in amount and moving from one type of storage to another. The only input and output is energy. Local systems by contrast are 'open', with both inputs and outputs independent.

MAJOR CHARACTERISTICS OF PRECIPITATION

Dew: Deposited at night on the ground surface - particularly on the vegetation canopy.

Fog-drip: Deposited from fog on vegetation and other obstacles. **Rime** is the frozen form.

Drizzle: Droplets less than 0.5 mm in diameter. Known as **freezing drizzle** when surface temperatures are less than 0°C.

Rain: Droplets more than 0.5 mm (usually 1-2 mm) in diameter. **Light rain** is less than 2 mm per hour. **Heavy rain** is more than 7 mm per hour.

Sleet: Partly melted snow, or a rain and snow mixture.

Snowflakes: Aggregations of ice crystals up to several centimetres across.

Snow grains (granular snow): Very small, flat, opaque grains of ice - the solid equivalent of drizzle.

Snow pellets (soft hail): Opaque pellets of ice, 2-5 mm in diameter, falling in showers.

Ice pellets (small hail): Clear ice encasing a snowflake or snow pellet.

Hail: Roughly spherical lumps of ice, 5-50 mm or more in diameter, showing a layered structure of opaque and clear ice in cross-section.

Component parts of the hydrological system

1. **Precipitation** is the major input. Generally speaking the more intense the input, such as in a storm, the shorter the duration. Precipitation formation depends on how the air is uplifted.

RAINFALL FORMATION

Convectional rainfall falls typically as showers or heavy downpours from cumulus or cumulo-nimbus clouds. Summer heating of the land surface sets up convection currents of warm air which rises, cools, and forms cloud droplets once the vapour condenses at the **dew point**. Dramatic cloud development may instigate thunder, lightning, and hail, but eventually the clouds themselves weaken the convection currents by reducing insolation.

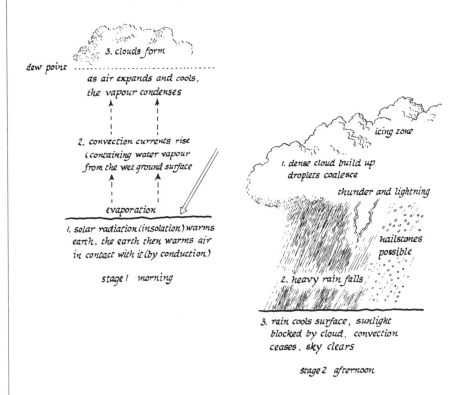

3. clouds form

dew point ....

as air expands and cools,
the vapour condenses

2. convection currents rise
(containing water vapour
from the wet ground surface

evaporation

1. solar radiation (insolation) warms
earth, the earth then warms air
in contact with it (by conduction)

stage 1   morning

icing zone

1. dense cloud build up
droplets coalesce

thunder and lightning

hailstones
possible

2. heavy rain falls

3. rain cools surface, sunlight
blocked by cloud, convection
ceases, sky clears

stage 2 afternoon

Relief (orographic) rainfall depends upon a physical barrier to deflect upwards the vapour-bearing wind. Clearly, whether or not clouds form, and rainfall results, depends upon the humidity of the rising air and the degree to which it cools.

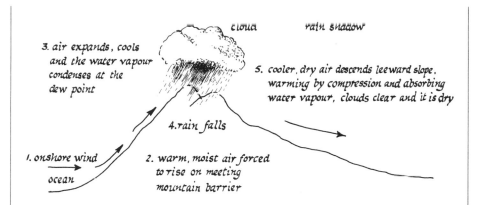

3. air expands, cools and the water vapour condenses at the dew point

cloud   rain shadow

5. cooler, dry air descends leeward slope, warming by compression and absorbing water vapour, clouds clear and it is dry

4. rain falls

1. onshore wind

ocean

2. warm, moist air forced to rise on meeting mountain barrier

Cyclonic (frontal) rainfall results from the horizontal convergence of air masses within areas of low pressure (a low pressure system). Warmer, moist, less dense air is forced to rise over colder, denser air - with cloud formation and precipitation resulting.

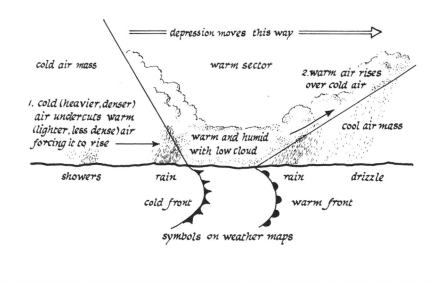

= depression moves this way =

cold air mass   warm sector   2. warm air rises over cold air

1. cold (heavier, denser) air undercuts warm (lighter, less dense) air forcing it to rise

warm and humid with low cloud

cool air mass

showers   rain   rain   drizzle

cold front   warm front

symbols on weather maps

Hydrologists require detailed information on not just the amount of precipitation, but its intensity, frequency, areal extent, and rate of movement. With sufficiently accurate knowledge, flood control systems, for example, can be planned more effectively. However, measurement is not yet globally standardised - with rain gauges, for example, prone to errors caused by rain splash and wind, especially if the precipitation falls as snow.

2. **Interception** affects the amount of precipitation to reach the ground. This depends upon the nature of vegetation cover and unnatural constructions. Temporary storage occurs on the interception surface until it evaporates, drips from foliage and roofs, and falls through between trees and buildings as **throughfall**, or trickles along branches and down tree trunks as **stemflow**.

The **interception capacity** of vegetation will vary over time. It is most effective when surfaces are dry and progressively decreases thereafter as the rainfall continues. Rainfall frequency and duration are, consequently, relevant.

Many variables determine how effective vegetation interception is. For example, the season may dictate leaf cover or its absence. Certainly, the type of vegetation cover could be critical in that conifers, for example, intercept more effectively through water clinging to their individual needles rather than coalescing and dripping off deciduous broad-leaves. Grassland and shrub vegetation will intercept less than woodland. Agricultural crops intercept even less, especially if bare soil is exposed between the rows.

3. **Evapotranspiration** is the combination of evaporation and transpiration. **Evaporation** is the diffusion of water to air (from liquid to gas). **Transpiration** is the loss of water through plant surfaces (mainly leaves). Plants have little control over it. Consequently they will continue to draw up more water from the soil, to replace losses, until there is not enough available. This is known as the **wilting point**. **Potential evapotranspiration** is the amount which would occur if there were unlimited water to be evaporated, such as from the surface of a lake. Warm, dry, windy conditions encourage the greatest rates of evapotranspiration - cold, moist, calm conditions, the least. Mixing air, by dry air moving in, increases it.

FACTORS INFLUENCING EVAPOTRANSPIRATION

1. **Temperature:** Solar radiation (insolation) heats the ground surface causing 'steaming'. Warm air can hold more moisture than when cold.

2. **Humidity:** As the saturation point is neared, evapotranspiration is inhibited.

3. **Wind:** Both speed and turbulence are relevant to the mixing of air and drying of surfaces, such as leaves.

4. **Vegetation cover:** Both shelter and shade will reduce evapotranspiration.

5. **Soil moisture content:** The opportunity for evapotranspiration decreases rapidly as the surface moisture content falls. Surface evaporation from a dry soil, for example, will be zero. Rainfall frequency is influential in this respect, for evapotranspiration is greater from a frequently wetted soil compared to one thoroughly soaked in a storm – even if the total precipitation is the same. This is because much of the storm water will go straight through – infiltrating, then percolating into the groundwater zone. In droughts or shallow, permeable soils the wilting point may be reached when the transpiration rate exceeds the uptake of soil moisture. Should evapotranspiration exceed precipitation, any surplus soil moisture will be utilised to leave a deficit. This is known as the **soil moisture deficit**.

6. **Soil texture:** This determines the amount of capillary action. Capillarity is more efficient in finer textures which retain more water. Sandy textures, for example, will dry out more rapidly than clay.

7. **Soil colour:** Dark soils absorb more heat, with greater evaporation in consequence. Lighter soils, by contrast, have a higher **albedo** (capacity to reflect light), so reducing evaporation.

8. **Depth of water table:** The deeper the water table, the less the evapotranspiration because there is less capillary action.

9. **Salinity of water:** There is less evapotranspiration as salinity increases.

10. **Depth of water in, for example, lakes:** Deep water has both less mixing and a lower surface area relative to depth – so reducing evapotranspiration. The opportunity for evaporation depends on the surface, which is why open stretches of water have 100 per cent opportunity, with vegetation and soil much less.

4. **Soil moisture** in the soil zone (the unsaturated zone above the water table) can flow laterally, close to the surface, as **interflow** or **throughflow**, percolate further down into the intermediate zone, or move upwards by capillary action, especially where evapotranspiration exceeds precipitation.

Water, therefore, entering the soil zone need not necessarily reach the zone of saturation. However, this is more likely on flood plains or river terraces than higher up the valley – not least because the capillary fringe (the depth of which depends on rock texture) is likely to extend well into the soil zone and even to the ground surface on these lower levels.

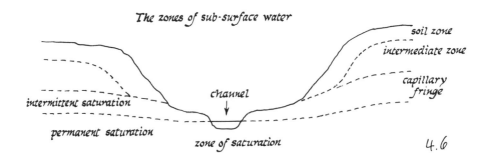

The zones of sub-surface water

soil zone

intermediate zone

capillary fringe

intermittent saturation

channel

permanent saturation

zone of saturation

4.6

Infiltration of water through the surface layers, into the soil zone, should not be confused with **percolation**, which must be regarded as downward movement to the water table. The **infiltration rate** is the rate of water absorption in millilitres per second or minute. **Infiltration capacity** is the maximum rate at which water can be absorbed by a soil in a given condition. Infiltration rates and capacity decrease with time as rain falls until a constant value is reached.

FACTORS INFLUENCING SOIL MOISTURE

1. **Soil type** and **solum (profile):** The porosity and structure of the soil is very influential in encouraging or restricting infiltration as, say, by indurated layers such as the iron pan in a podsol.

2. **Soil surface conditions:** Infiltration will be restricted by compaction, frost, or steep slopes – all encouraging surface runoff (overland flow). Desiccation (sun) cracks, deep ploughing, or flat relief would, conversely, increase infiltration. (See also *antecedent rainfall* below.)

3. **Surface cover:** Vegetation cover has already been demonstrated to be influential in interception, especially woodland with undergrowth acting as **secondary interception** beneath. Consequently, soil moisture content is affected, although not necessarily reduced, in that evaporation, for example, may be lessened by the shade. The roots of vegetation, by their penetration into the soil, make infiltration easier.

4. **Amount of antecedent rainfall:** Earlier prolonged rainfall may have already saturated the soil, so prohibiting further infiltration. However, after a long dry spell heavy rain may lie on the surface for some time before the upper layers become wet and soft enough to absorb the moisture.

5. **Groundwater** is water filling all pore spaces above impermeable rocks in the zone of saturation. It is rich in dissolved mineral salts and can flow laterally as **groundwater flow** or **baseflow**. Meteoric, connate, and juvenile sources are differentiated. **Meteoric** water is precipitation which has percolated down. **Connate** water, in minute amounts, will have been trapped in the rocks when they were formed. **Juvenile** water is from volcanic sources reaching the upper layers for the first time. However, only a small proportion of the groundwater zone consists of **aquifers** – rocks which store and transport significant amounts of water, hence the surprising reality that *'groundwater is mostly rocks'*.

Water can only occupy the pore spaces (**interstices**) between solid rock particles or fragments. The size and shape of any pores will depend upon the geological origins of the material. Shale and Millstone Grit, for example, are relatively impermeable compared to loosely compacted sandstone.

THE GROUNDWATER BALANCE EQUATION

Changes in storage = recharge − discharge

$$\Delta S \qquad = \qquad Qr \qquad - \qquad Qd$$

Recharge = percolating precipitation + influent seepage through river beds and banks + leakage from nearby aquifers + artificial sources such as reservoirs and irrigation.

Discharge = evaporation loss from the surface (if the water table is high enough) + effluent seepage (to the sea) + leakage to other aquifers + artificial abstraction (from wells and boreholes).

6. **Runoff** includes channel flow (streamflow) and overland flow. **Streamflow** refers to water in permanent channels. **Overland flow** via tiny rills or, if the ground is particularly compacted, as sheetflow, occurs when the rate of precipitation exceeds the rate at which the soil can absorb it. The collective term runoff is the excess of precipitation over evapotranspiration once an allowance has been made for storage. This is expressed in the water balance equation. Indeed, runoff is more regular than precipitation because the storage factor smooths out irregularities.

THE WATER BALANCE EQUATION

Runoff = precipitation − (evapotranspiration ± changes in storage)

$$Q = P - (E \pm \Delta S)$$

Clearly the storage capacity must be filled before runoff occurs. Runoff is then influenced by climate, the catchment, and human activities.

Climate: Meteorological factors are dynamic and dictate the balance between precipitation input (type, quantity, intensity, duration, and occurrence of antecedent rainfall) and evapotranspiration (influenced by temperature changes, vegetation, and wind).

Catchment: These factors are mainly static and include area, lithology/geology, slope, altitude, soil type, vegetation, and drainage network density.

Human activities: Human factors include irrigation, drainage, deforestation, urbanisation, and the building of dams and aqueducts, for example. The influence of human activity should never be underestimated.

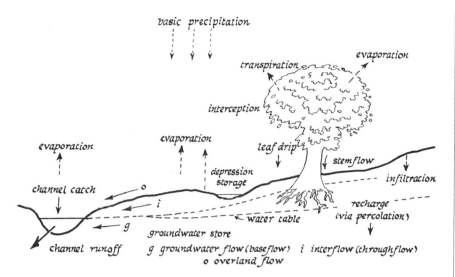

basic precipitation

transpiration

evaporation

interception

evaporation

evaporation

leaf drip

stemflow

infiltration

channel catch

depression
storage

recharge
(via percolation)

channel runoff

water table

groundwater store

g groundwater flow (baseflow) i interflow (throughflow)
o overland flow

Factors influencing runoff

Channel precipitation: 'Channel catch' is a relatively small amount because the channel area, at most, will only be 5 per cent of the total catchment area.

Overland flow: Again, this is very small because most will infiltrate unless the ground is bare rock, frozen, or concreted, for example.

Interflow (throughflow): This may account for up to 85 per cent of the total runoff. It may be rapid and immediate if the ground is already wet, or delayed if not.

Groundwater flow (baseflow): This is very slow and may lag behind precipitation by weeks or even years. The main long-term component of runoff, it is especially important during periods of drought.

Snowmelt: If melting is slow then the water resulting will percolate to the groundwater. However, if melting is rapid, the water will act like storm runoff, especially if the soil is still frozen.

Stream types are determined by the relationship between baseflow (of groundwater) and quickflow (which is direct runoff). Most rivers have ephemeral and intermittent streams feeding their upper reaches, depending on the time of the year. **Ephemeral streams** result from quickflow only after a storm or rapid snow melt. Rills are a good example with the water table always below the bed. **Intermittent streams** will flow in wet periods but dry up during droughts. The water table may rise to the level of the bed, so allowing groundwater to make a contribution. **Perennial streams** would include most rivers because the water table is always above the bed. This allows water to enter through both bed and banks. Groundwater, consequently, makes a permanent contribution.

River regimes

The regime of a river refers to seasonal variations in runoff. An equatorial river, for example, might be expected to show even flow throughout the year compared to marked variations between the wet and dry seasons in tropical ones.

Simple regimes are differentiated from complex ones. A **simple regime**, for example, demonstrates one peak period of runoff – as in a monsoon climate. **Complex regimes**, in contrast, have more than one peak. For example, a river in continental Europe might display a runoff maximum following spring snow melt, followed by a later peak reflecting convectional summer rain. A river with tributaries rising in different climatic zones would complicate a discharge pattern further.

It is of note, however, that most rivers are likely to have a simple regime near their source, but complex nearer their mouth. These regimes may be shown on **long period hydrographs** – with the water year starting in October when storage, in Britain, is likely to be zero.

Given that few rivers, nowadays, are wholly natural, one can expect a certain degree of regulation, such as through drainage basin[1] management – not least water abstraction for agricultural, industrial, and domestic use. River discharge can change indirectly too – through, for example,

---

[1] A **drainage basin** (**catchment area**) is the area of land drained by a river and its tributaries. Its perimeter of **watersheds** separate adjacent basins – and unless a **basin of inland drainage**, may also include a stretch of coast.

deforestation or urban development reducing natural infiltration of precipitation and accelerating overland and artificial drainage flow to river channels.

However, effective management of discharge requires more detailed information than the annual record of the long period hydrograph. **Short period (storm) hydrographs** demonstrate the river's response to specific rainfall events. Given the susceptibility of some urban areas to flash floods, detailed knowledge of discharge behaviour is essential if effective levées, culverts, and so on are to be engineered.

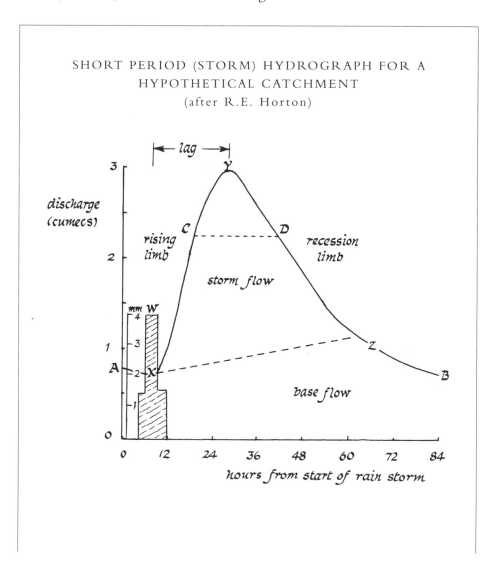

SHORT PERIOD (STORM) HYDROGRAPH FOR A
HYPOTHETICAL CATCHMENT
(after R.E. Horton)

AX   The gradual reduction of groundwater flow (baseflow) since the last precipitation.

W   The peak of the storm (precipitation input).

X   Discharge starts to increase as the precipitation, initially via surface runoff, eventually reaches the channel.

XY   The **rising limb** shows the rapid increase in discharge after the catchment infiltration capacity has been reached.

Y   The point of peak flow.

WY   The time between peak precipitation and peak discharge. This is known as the **lag time**.

YZ   Discharge decreases as minor channels drain dry in the **recession limb**.

ZB   End of storm runoff with mainly groundwater flow (baseflow) again.

XZ   The **time base** from start to finish of the storm runoff.

CD   Bankfull discharge level above which the flood plain is inundated.

Variations in hydrograph shape from river to river demonstrate not just human influence but physical characteristics of the catchment area too. A very sharp peak, for example, results from high immediate surface runoff with little absorption and storage of water in the drainage basin due to, for example, impermeable bedrock and steep relief.

CATCHMENT EXAMPLES

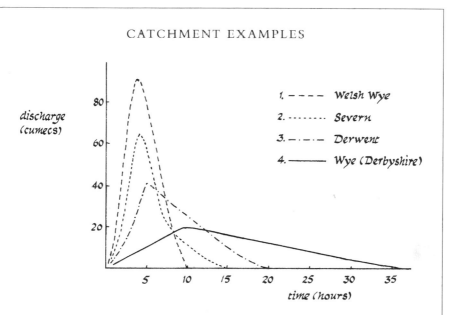

River 1 has an impermeable catchment with mainly moorland vegetation. Runoff is 83 per cent of input.

River 2 has an impermeable catchment with mainly coniferous forest. Runoff is 64.4 per cent of input.

River 3 has a semi-permeable catchment of sandstone and shale with a mixture of woodland and agricultural land. Runoff is 77 per cent of input.

River 4 has a permeable Carboniferous limestone catchment and is agricultural land. Runoff is 70 per cent of input.

Drainage basin morphometry

Morphometry (the measurement of shape of any natural form be it animal, plant, or drainage basin) allows quantitative rather than purely subjective study of comparative catchment areas. Linear, areal, and relief elements of fluvial morphometry are identified. **Linear properties** of the stream channel are one-dimensional and include stream orders and lengths. **Areal properties** are in two dimensions and include the surface area and shape of basins. **Relief properties** add the third dimension of channel and ground surface gradients.

1. **Linear properties**:

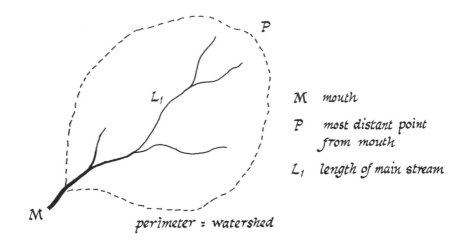

M   *mouth*

P   *most distant point from mouth*

L_1 *length of main stream*

perimeter = watershed

Definition of the perimeter is usually quite easy because watersheds are easily recognised. Groundwater divides usually coincide with these.

Network definition is far more difficult because the source may be boggy ground or ephemeral rills. Also the sources will change with headward erosion, infilling, channel migration, and storms.

Stream order (as devised by R.E. Horton and subsequently modified by A.N. Strahler) divides each network into channel segments which are numbered according to a hierarchy of orders of magnitude. Each 'finger tip' channel is designated **first order**. At the junction of two first order streams a **second order** segment starts. Where two second order streams meet a **third order** segment starts – and so on. The **trunk stream** carries the highest order for the system and the whole drainage basin is described in terms of this highest order stream.

1st order	10
2nd order	4
3rd order	2
4th order	1

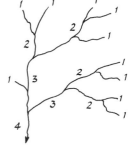

a 4th order basin

THE BIFURCATION RATIO

Bifurcation means forking or branching. The **bifurcation ratio (Rb)** simply measures how much branching takes place. It exists between the number of segments (**N**) of any given order (**u**) in relation to the number of segments of the next highest order (**u+1**). It indicates, therefore, how many streams of one order, on average, are required to produce a stream of the next highest order.

$$\mathbf{Rb} \ = \ \frac{\mathbf{Nu}}{\mathbf{N(u+1)}}$$

Example:

Order	Number of streams/segments		Rb
1st	110		
2nd	27	1st: 2nd	$\frac{110}{27} = 4.1$
3rd	5	2nd: 3rd	$\frac{27}{5} = 5.4$
4th	2	3rd: 4th	$\frac{5}{2} = 2.5$
5th	1	4th: 5th	$\frac{2}{1} = 2.0$

$$\text{Mean } \mathbf{Rb} \ = \ \frac{4.1 + 5.4 + 2.5 + 2.0}{4} \ = \ 3.5$$

Mean values between 3 and 5 are characteristic of humid climatic systems such as Britain's. However, structural effects can elongate a drainage basin, so increasing the ratio dramatically. Calculation of the bifurcation ratio is an important factor when determining the peak runoff point on a storm hydrograph. Indeed, the higher the ratio, the more likely the drainage network will flood.

$Rb = \frac{9}{1} = 9$

2. **Areal properties**: An important purpose of fluvial morphometry is to obtain quantitative data about fluvial geometry and to correlate it with hydrological information. The relationship between discharge and drainage basin area, for example, can be stated objectively. Clearly, as area increases so does discharge. But hydrologists could estimate the discharge by calculating the catchment area above a required gauging point – so gaining essential data in order to enable the safe design of bridges, dams, irrigation schemes, and so on. Another obvious relationship is between drainage density and basin area.

$$\textbf{Drainage density} \;=\; \frac{\textbf{total length of all streams (km)}}{\textbf{area of basin (km}^2\textbf{)}}$$

This is expressed as km per km^2 – the lowest figures occurring on permeable rocks in regions of low rainfall. Indeed, permeability is critical in that sands and gravels allow high infiltration and so low runoff. Clays and shales, by contrast, demonstrate high drainage densities. Resistance is also relevant in that hard rocks like granite and gneiss have low drainage densities because channels are so difficult to cut. Vegetation cover is also influential given the high infiltration rates associated with good forest or grass cover in humid climates – hence lower drainage densities than on cleared land.

Drainage densities affect patterns of runoff. Surface runoff is rapidly removed in basins with high drainage densities. Consequently, there is a shorter lag time between precipitation and peak flood, which is likely to be higher on the hydrograph.

Finally, the shape of the drainage basin tends to have little influence on drainage density, unless there is strong structural control, such as in California where elongated basins have high relief and steep slopes.

In summary, therefore, the factors affecting drainage basins are rock type (permeability and resistance), vegetation, and basin shape.

3. **Relief properties**: The gradient of the stream channel (its **long profile**) is affected by many variables including discharge, velocity, depth, width, quantity, and calibre of load. Long profiles are usually concave upwards because larger channels downstream are more efficient and so require less gradient. Indeed, the slope of valley sides complement those of the stream channel. Steep slopes accelerate runoff whilst gentle ones reduce

its rate. Consequently, the mean basin slope affects the storm hydrograph and relief controls the rate of sediment loss.

HORTON'S LAWS OF DRAINAGE COMPOSITION

Horton's pioneering work on morphometric techniques in the mid-1940s has done much to advance more objective approaches to hydrology. The four laws cover not just the linear properties of drainage basins, but areal and relief aspects too.

1. **The law of stream numbers** states that '*the number of stream segments of successively lower orders tend to form a geometric series, beginning with a single segment of the highest order, and increasing according to a constant bifurcation ratio.*' Stream number is, therefore, related to stream order by geometric relationship. There are more first order streams than second, more second order than third - and so on.

If **Rb** = 3 and the trunk segment is sixth order then there will be

- 1   sixth order
- 3   fifth order (3 x 1)
- 9   fourth order (3 x 3)
- 27 third order (3 x 9)
- 81 second order (3 x 27)
- 243 first order (3 x 81)

This may be expressed mathematically as $\mathbf{Nu = Rb^{(k-u)}}$ where **Nu** is the number of segments, **k** is the order of the trunk stream, and **u** is the order. Hence the number of fourth order segments in the example above would be $3^{(6-4)} = 9$.

Results can then be plotted on semi-log graph paper (the vertical scale in log cycles, the horizontal normal arithmetic) to obtain a virtually straight line. As stream order increases the number decreases - a negative correlation.

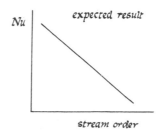

2. **The law of stream lengths** states that *'the average lengths of streams of each of the different orders in a drainage basin form a geometric series.'* First order streams are shorter than second order streams and as the order increases so does the average length. This can be done for each segment, or by adding first order average (mean) to second order mean and so on to produce a cumulative figure. Horton found this cumulative method to be more accurate, stating that *'the cumulative mean lengths of stream segments of successive orders tend to form a geometric series, beginning with the mean length of the first order segments and increasing according to a constant length ratio.'*

Again, results plotted on semi-log graph paper for stream order against cumulative length will produce a straight line. As stream order increases so does mean length – a positive correlation.

Example:

Order	Number	Mean length (km)	Cumulative mean length (km)	
1	39	0.56	0.56	
2	9	0.93	1.49	(0.56+0.93)
3	3	3.38	4.87	(1.49+3.38)
4	1	5.49	10.36	(4.87+5.49)

The length ratio ($\mathbf{R_L}$) can be calculated between successive orders by

$$\mathbf{R_L} = \frac{\mathbf{Lu}}{\mathbf{L\,(u-1)}}$$

where **Lu** is the mean length of all segments of order **u**

$\mathbf{R_L}$ between second and third order streams in this network would, therefore, be

$$\frac{3.38}{0.93} = 3.63$$

3. **The law of drainage basin area** states that as stream order increases, so does the area of the drainage basin.

Example:

Order	Number	Area (km²)
1	39	0.404
2	9	0.602
3	3	8.583
4	1	32.639

The areas are calculated by superimposing a tracing of the catchments (at a known scale) onto graph paper - then simply counting the squares covered.

Plotting the results on semi-log graph paper produces another straight line 'geometric relationship'. As stream order increases so does drainage basin area - a positive correlation.

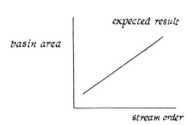

4. **The law of stream gradient** states that low order streams have steep gradients and high order streams have gentle gradients.

Example:

Order	Number	Gradient (m km⁻¹)
1	39	71
2	9	53
3	3	29
4	1	20

Gradient is averaged for all streams of that order.

$$\text{Gradient} = \frac{\text{fall in height}}{\text{total length}}$$

Once again, results plotted on semi-log graph paper produce a straight line. As stream order increases, slope decreases – a negative correlation.

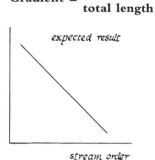

5
RIVERS

Rivers are fundamental in the shaping of many landscapes. Not only do they **erode**, but also **transport** and **deposit** the products of weathering and mass movement. Individual stream channels[1] not only provide water, but also remove excess – therefore they are **transport systems** of water and sediment.

When we analyse stream channels, the role of slopes is fundamental in that changes in either will affect the other.

As illustrated in *Hydrology*, water reaches the channel through direct (channel catch) precipitation falling on the river, overland (surface) flow from rainwash and rills plus interflow (throughflow) and groundwater flow (baseflow) beneath the surface entering through beds and banks.

Sediment comes from adjacent valley slopes and reaches the channel by mass movement or surface (sheet) wash. Some will come from collapsing river banks.

River energy

Potential (stored) energy is fixed by the altitude at which the stream rises.

Kinetic energy varies along the course of the stream according to discharge (volume of water in the channel) and slope (channel gradient) – both of which determine the velocity (speed of flow).

All fluvial processes depend on the amount of energy available. This is used to overcome friction with bed and banks, transport the load, and erode.

This is all in delicate balance. For example, different parts of a river may display **erosion** where there is excess energy after transporting the load, **deposition** where energy is insufficient to move the load, or **equilibrium** where transport energy of around 10 per cent matches

---

[1] The term **stream channel** includes channels of all sizes and permanence.

the material supply – leaving around 90 per cent to be dispersed as
frictional heat. Channels, consequently, adjust in shape and size to
accommodate changes in the amount of water and sediment.

Discharge = width x depth x velocity
(cumecs)   (cross-section in m^2) (ms^{-1})
Q = w d v

Therefore, as discharge increases downstream so must width, depth, and
velocity. The **turbulent flow** of mountain torrents may look faster, but
this is due to extreme disruption of water on a rough bed. Maximum
velocities may be greater, but the average velocity is slower, with so much
energy expended overcoming friction. **Laminar flow** (with one layer
sliding smoothly over another at different speeds) accounts for the higher
velocities downstream.

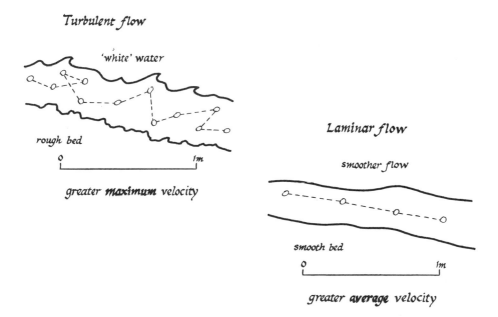

Turbulent flow

'white' water

rough bed

0   1m

*greater **maximum** velocity*

Laminar flow

smoother flow

smooth bed

0   1m

*greater **average** velocity*

Channel processes: transport, erosion, and deposition

As stated earlier, rivers are transport systems for water and the sediment
load.

◆ **Suspension load** represents solid particles carried within the current, but not touching the bed. It can be measured by collecting a sample, drying and then weighing it.

◆ **Bedload** represents solid particles moved along the bed – rolled as **traction load** or bouncing as **saltation load**. This is difficult to measure because the traps required alter the current. However, it is usually about 10 per cent of the suspension load and much affected by bed roughness.

◆ **Dissolved load** is in solution and so can only be measured with lengthy and expensive chemical analysis. It could, however, represent half the total in a stream without much bedload.

◆ **Flotation load** comprising surface twigs, leaves, and so on is usually included in the suspension load.

The ability of a stream to carry its load is called its **efficiency**. However, the conditions for efficient transport of bedload will not be the same for mainly dissolved load.

$$\textbf{Efficiency} \; = \; \frac{\textbf{capacity (grams s}^{-1}\textbf{)}}{\textbf{discharge (cumecs)}} \; \textbf{x \% slope of channel bed}$$

Capacity represents the maximum load of sediment of all types, irrespective of particle size, that the current can transport. It increases with discharge and/or gradient, but decreases when the calibre of the particles increases. As velocity increases, therefore, so larger material can be moved – as shown on the **Hjulström curve**.

Competence refers to the maximum size of particle that the stream is capable of transporting. Indeed, the maximum particle mass that can be moved increases with the sixth power of the velocity. Hence, for example, a doubling of velocity could lift a particle 2^6 (2x2x2x2x2x2) = 64 times the mass of before. Competence is determined by finding the largest boulder which the stream has definitely moved at some time.

THE HJULSTRÖM CURVE

This shows the relationship between stream velocity and particle size, plotted on logarithmic scales.

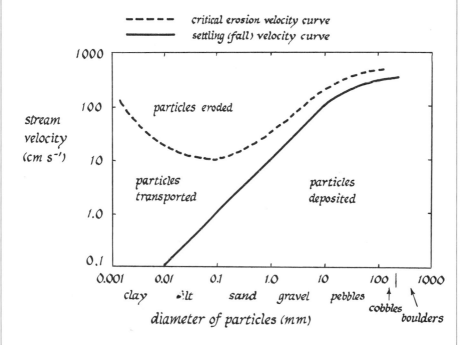

The boundary between erosion and transport is known as the **critical erosion velocity curve** and gives the approximate velocity needed to pick up and transport in suspension particles of various sizes. Material on the bed is inert and so requires far more energy from the moving water to **entrain** (pick it up) than that needed to subsequently keep it moving. This is because the particles' momentum helps keep them going. This material is then responsible for most of the subsequent erosion – providing there is sufficient extra energy from higher velocities to allow this, for transport clearly requires much less energy than erosion. The lower **settling (fall) velocity curve** shows the velocities at which particles of a given size become too heavy for transportation – so falling out of suspension to be deposited. Notice how small silt and clay particles require disproportionately larger velocities to raise them from the channel bed. This is both because of their cohesion and also their limited resistance to water flow on the channel bed – so requiring a more energetic stream to lift them.

Erosion occurs whenever there is an excess of energy remaining after friction is overcome and the load is transported. There are four main processes:

1. **Abrasion (corrasion)** is the wearing away of bed and banks by the load acting like sandpaper. Pebbles caught in swirling currents, for example, will abrade **pot-holes**.

2. **Attrition** is the reduction in size of the particles themselves – caused by them impacting with each other.

3. **Solution (corrosion)** is independent of both discharge and velocity and, therefore, a purely chemical action related to acids in the water.

4. **Hydraulic action** is the mechanical loosening and removal of particles by water pressure alone – increasing air pressure in any cracks which will consequently expand and weaken. **Cavitation** in turbulent flow is also a form of hydraulic action whereby air bubbles implode – with the resulting shock waves weakening the banks. It is particularly evident where water pressure decreases as a river 'shoots' rapids and waterfalls.

All erosion is most effective in flood conditions. Banks are more vulnerable than beds, especially to undercutting which leads to collapse. Indeed the banks may collapse after a flood because the supporting pressure of water has been released. Certainly **bank lithology** is important because permeable materials, such as sands and gravels, can become waterlogged and so prone to collapse. Clays by contrast are quite coherent and so can support steeper banks before collapsing.

Deposition occurs when there is insufficient energy for transport. (Since deposition and entrainment can occur at the same time, we normally accept this to be a net state of more deposited than removed.) The heaviest (largest) particles are always dropped first as velocity is reduced – due to gradient reduction on reaching the valley floor or a **base level** such as the sea or a lake, or discharge reduction due to drought, evaporation, percolation, flood water subsidence, or abstraction.

Channel geometry: cross-section, long profile, and plan

Channel geometry (**form**) represents a classic 'chicken and egg' situation in that form determines flow, yet flow determines form!

The **cross-section's** relation between depth and width is critical. Indeed, the form is related to the main function of the channel. For example, a large bedload is likely to be associated with a wide and shallow cross-section, contrasting with a predominantly suspended load in a deep, narrow channel. This is due to the **friction index** of water contact between bed and banks, calculated by measuring the **wetted perimeter** (distance across banks and bed measured using a chain).

Both channels carry 16 m³ over a distance of 1 m. However, B's wetted perimeter of 12 m has a lower friction index than A, with a wetted perimeter of 18 m. B's lower friction index enables water to flow faster in the centre of the stream - efficiently carrying load in suspension. A's high friction index, however, reduces the velocity at bed and banks - so concentrating the energy to move bedload efficiently.

Comparing channels of different cross-section is achieved by calculating the **hydraulic radius.**

$$\textbf{Hydraulic radius} \ = \ \frac{\textbf{volume index (cross-section area)}}{\textbf{friction index (wetted perimeter)}}$$

$$A = \frac{16}{18} = 0.89 \qquad B = \frac{16}{12} = 1.33$$

NB: For ease of calculation we always take the volume index to be cross-sectional area.

A hydraulic radius less than 1 is low. Basically, the higher the value, the more efficient the channel at transporting suspended load.

Clearly, we can apply this formula to prediction of channels carrying different volumes of water:

Model gorge

4m   bankfull discharge

2m   normal discharge

3m

	Normal	Flood
Cross-section area	6 m²	18 m²
Wetted perimeter	7 m	15 m
Hydraulic radius	$\frac{6}{7} = 0.86$	$\frac{18}{15} = 1.20$

We can conclude that the model gorge at bankfull, prior to flood, experiences less friction than normally and so is more efficient. Suspended load is likely, therefore, to be greater prior to flooding, as the colour of any river in spate will confirm.

Model river with flood plain

flood discharge

5m

10.5m   2m   2m   10.5m

normal discharge   3m

	Normal	Flood
Cross-section area	6 m²	18 m²
Wetted perimeter	7 m	29 m
Hydraulic radius	$\frac{6}{7} = 0.86$	$\frac{18}{29} = 0.62$

The channel becomes less efficient, however, once the river floods. Deposition, therefore, is likely to follow.

Even if channels have the same shape, their size will affect efficiency in that bigger channels are more efficient.

The river's **long profile** can be drawn from source to mouth to show changes in altitude and so its bed-slope (gradient). The mouth, usually the sea, forms the ultimate base level where potential energy reaches zero. However, the long profile may be interrupted by numerous base levels such as bands of hard rock or lakes.

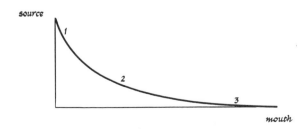

source

1

2

3

mouth

1. The **upper course** is characterised by a gradient of 1:10 or steeper with a V-shaped valley cross-profile winding around interlocking spurs. A wide shallow channel with low hydraulic radius is probable. Therefore, erosion by downcutting is dominant - resulting in a high percentage of large calibre bedload being transported.

2. The **middle course** should have a gradient between 1:10 and 1:100 and an open valley cross-profile. A deeper channel with higher hydraulic radius is likely. Erosion by lateral action dominates - so developing meanders. A low proportion of smaller calibre bedload is transported, but a high percentage of sediment in suspension.

3. The **lower course** should have a gradient gentler than 1:100 with a wide flood plain. A large deep channel with high hydraulic radius and exaggerated meanders is usual. Deposition dominates, especially in flood, with all sediment of small calibre, transported mainly in suspension.

Clearly these are generalisations, for along the long profile the gradient, cross-section, and dominant processes can alter from one reach to another. Observation, recording, and analysis is, therefore, needed at each point.

'Slope is an inverse function of discharge.' Discharge, therefore, increases from source to mouth as slope decreases. This is because large channels have high hydraulic radii - so enabling them to maintain velocity on gentler gradients.

Grade refers to a state of **dynamic equilibrium** whereby a river (or, similarly, a slope) achieves a balance between erosion, transport, and deposition. In a graded reach - more likely than along the entire long profile - the slope of the channel is in a **steady state** with no net gain or loss of sediment. All factors are in balance with all kinetic energy used to transport the water and sediment load supplied - without excess for erosion or deficiency for deposition. Channels, however, are rarely permanently stable and will adjust their form to suit the environmental conditions of the day, season, or era. Since grade, therefore, represents a state of such delicate balance, perhaps it should only be applied to individual reaches rather than the whole stream. Grade can, however, theoretically be thought of in the long-term, such as in the concave sloping river long profile illustrated earlier, whereas steady state is always a short-term phenomenon.

But how does this dynamic equilibrium operate? Just as slopes, theoretically, maintain the most efficient angle, so channels, likewise, maintain an efficient gradient by self-regulating processes. Three independent variables are involved – discharge, load, and ultimate base level. Other variables are interdependent – channel width and depth, the nature of load, and the channel's tendency to meander or braid. **Negative feedback** operates whereby one factor going out of balance sets off a chain reaction which eventually leads back to equilibrium:

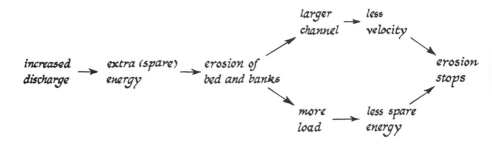

The river in **plan** shows the sinuosity of the channel and its tendency to meander, braid, or, in rare cases, run straight.

$$\textbf{Sinuosity} = \frac{\textbf{channel length}}{\textbf{straight-line distance}}$$

NB: The nearer the value to one, the straighter the channel.

Straight channels, unless canalised, will rarely last for more than ten times their width and usually comprise a sequence of shallow wider **riffles** and deeper narrower **pools** – usually five to seven times the channel width apart.

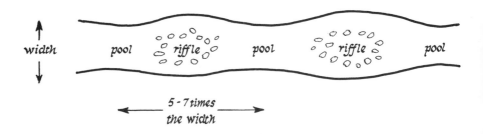

Meandering channels also display mathematical regularity - with riffles on the straights, pools on the bends, and a wavelength around ten times the width.

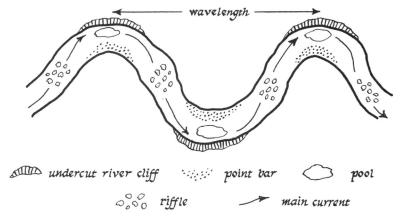

⟨undercut river cliff symbol⟩ *undercut river cliff*   ⟨point bar symbol⟩ *point bar*   ⟨pool symbol⟩ *pool*

⟨riffle symbol⟩ *riffle*   ⟨main current symbol⟩ *main current*

When rivers meander their average gradient is reduced because of the extra channel distance covered by winding.

If B was 5m lower than A the gradient would be 5/100 = 0.05. However, on meandering the channel is now 150m long, so reducing the average gradient to 5/150 = 0.033.

A ——————— *100m* ——————— B

A ⟨wavy line⟩ *150m* ⟨wavy line⟩ B

$$\text{Sinuosity} = \frac{150 \text{ m}}{100 \text{ m}} = 1.5$$

Gradient reduction in this manner can help the river to maintain its equilibrium by slowing down the velocity. The excess energy is more easily absorbed over the increased distance. Lateral erosion concentrating on the banks of the bends, however, predominates over vertical erosion of the bed.

Braided channels, in contrast to meandering ones, cope better with fluctuations in discharge because when it becomes too great the channel splits into many smaller ones - so increasing friction, reducing velocity, and slowing down the flood.

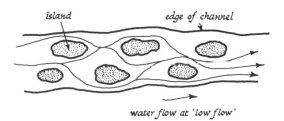

water flow at 'low flow'

The relationship between meandering and braiding is very sensitive 'on the line', which can result in rapid changes to channel form, with only minor alterations to slope or discharge. Clearly rivers braid to accommodate more water and/or move lots of bedload in the steep upper reaches – where wide shallow streams need friction to move high calibre material. Rivers in monsoon areas demonstrate rapid adjustments to channel form in response to the variable discharge associated with this climate.

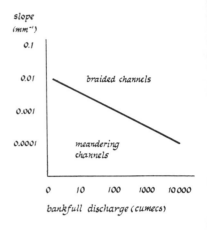

Rivers cannot change their catchments and are controlled, therefore, by the inputs of water and sediment supplied to them. However, they can adapt their geometry to cope with any changes in input. The channel plan, for example, adapts by braiding to cope with increased discharge or by meandering when the gradient decreases. Similarly the channel cross-section may adapt through bank collapse or bed scouring which would also, with deposition elsewhere, change the long profile.

To take our knowledge further, particularly of channel features, we need to have confidence in measuring the processes determining them.

Quantitative analysis, further channel processes, and resulting features

Width and **depth** can be measured accurately with the familiar tape measure, metre rule and/or ranging pole, so allowing cross-sections to be drawn and averaged if required.

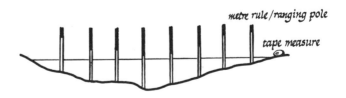

Velocity, however, is less straightforward – normally involving floats or current meters. Should such methods prove impractical, such as during a flood, Manning's 'n' is available.

CALCULATING VELOCITY USING MANNING'S 'N'

Manning's 'n' is a channel roughness coefficient whereby the higher the value, the rougher the bed and banks. The coefficient may be read from published tables.

$$\textbf{Velocity} \; = \; \frac{\textbf{slope x hydraulic radius}}{\textbf{Manning's 'n'}}$$

◆ Floats are restricted to surface velocities – their progress timed and averaged over set distances – providing they do not get stuck. However, they must be biodegradable and necessitate patient colleagues!

◆ Current meters (flow vanes) may be expensive, but their accuracy and capacity to operate at varying predetermined depths makes them essential for serious research.

Since friction against the stream bed retards the flow – a drag effect transmitted upward – the resulting **velocity profile** high-lights the overestimate inevitable if surface velocity, alone, is measured. A 20 per cent reduction of this value is, therefore, necessary.

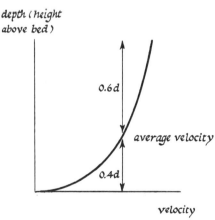

Average velocity is measured at 0.4 of the depth from the channel bed.

Clearly, velocities measured at varying depths across the channel allows us to interpolate **isovels** (lines of equal velocity) which invariably illustrate the retarding effect of bank and bed friction. Also the main current, which follows the deepest channel (or **thalweg**), could be plotted accurately in meander plans. This would help explain the incidence of erosion and resulting undercut **river cliff** where energy is concentrated, and the deposition of **point bars** in the slack waters of the inside (opposite) bend. Accurate measurement allows us to monitor and understand changes over time.

Vertical erosion (downcutting) dominates in the upper course of a river. Aided by turbulence, the bed is lowered by abrasion and enlarging pot-holes coalescing. V-shaped valleys and gorges, interlocking spurs, and waterfalls and their remnant rapids are all features which result.

Lateral erosion dominates in the middle course where the gradient is less steep, bed conditions are smoother, and the calibre of load smaller. Channel banks are undercut as the main current swings by centrifugal force to concentrate energy on the outside bend, with river cliff. Lateral turbulence, in reality caused by a 'corkscrewing' **helicoidal** flowing cross-current, ensures that material eroded from the concave outer bank, is deposited on the convex inside bank of the next bend, with point bar.

Migrating meanders, developing a flood plain bordered by the eroded ends of spurs called **bluffs**, and **river terraces** of various heights, are all characteristic features.

Deposition dominating in the lower course, where gradient is gentle, the bed smooth, and the calibre of load very small, will again result in distinguishing channel features. Variations in available energy may allow entrainment as well as the deposition, but rarely erosion – even in flood conditions, when the alluvium absorbs excess energy by being reworked. The flood plain is likely to be wider than the meander belt, with **ox-bow lakes**, braiding, and a raised river bed with **levées**, often now strengthened to reduce further flood danger. **Delta** formation at the coast is the ultimate depositional feature.

Drainage patterns

Drainage patterns depend on the original surface gradient, the lithology of the catchment, and the water supply. Five patterns in plan are recognised:

1. **Radial** patterns flow outwards like the spokes on a wheel. Volcanic cones and domes of homogeneous rocks, such as in northern Arran, demonstrate this pattern.

2. **Dendritic** patterns branch and
subdivide like the veins on a leaf.
Tilting homogeneous rocks encourage
this pattern as illustrated by the River
Tamar and its tributaries near Plymouth.

3. **Trellised** patterns display right angles
where alternate beds of hard and soft
rocks outcrop at the surface. Scarp and
vale topography, such as in the South
Downs, illustrates this well.

4. **Parallel (rectilinear)** patterns are very
similar to trellised, albeit demonstrating
exaggerated elongation along the straight
slopes of generally homogeneous rocks,
such as in Cleveland.

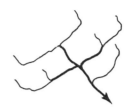

5. **Deranged** patterns are **confused** and so
impossible to classify as any of the above.
Gentle, irregular slopes – especially if recently
deglaciated – may display this. Rannoch
Moor shows this pattern.

Consequent and inconsequent drainage patterns

Consequent (accordant) drainage patterns are still directly related to,
and controlled by, the rocks and slopes over which they flow. Trellised and
parallel patterns result – and are often prone to river capture.

Inconsequent (discordant) drainage patterns no longer relate to the
rocks and slope over which they flow because they maintain the pattern
they developed originally. **Superimposed** drainage patterns in the Lake
District are a good example, whereby the original pattern developed on
exposed rocks which were then eroded away by rivers (and, latterly, ice)
which have since maintained their pattern on the underlying rocks.
Another example is **antecedent** drainage which occurs during and after

isostatic uplift of the land, when the river continues vertical erosion fast enough to maintain its course. Gorges such as the Arun in the Himalayas provide good examples, with vertical downcutting keeping pace with the orogenesis.

INSEQUENT, CONSEQUENT, SUBSEQUENT, AND OBSEQUENT STREAMS

Insequent streams (**I**) follow courses unrelated to the underlying rock. **Consequent** streams (**C**), however, follow a course determined by the original **dip** of the underlying rock, such as with the main river in a trellised pattern. **Subsequent** streams (**S**) develop at right angles to the dip following the **strike**, such as the tributaries on softer rocks. Finally, **obsequent** streams flow in the opposite direction to the consequent, such as the tributary labelled **O**.

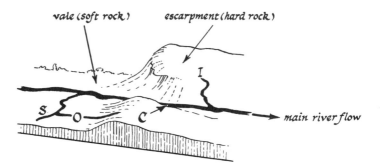

NB. The main river has cut through the hard rock escarpment to produce a 'river gap'

River capture

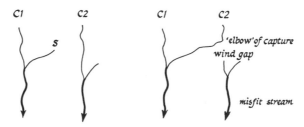

This occurs when a rapidly eroding subsequent river captures the headwaters of a consequent one. The subsequent tributary's rapid downward and headward erosion is normally due to flowing over softer rocks. Once interception has occurred a **misfit stream**, with reduced volume, is left below the **wind** or **dry gap** marking the **elbow of capture**. A consequent river can become progressively more dominant by capturing several parallel rivers. The North Tyne's capture of the Blyth, Wansbeck, and Rede in Northumberland is a classic example of this.

Changes in base level

Base level, as stated earlier, is the lowest point to which erosion by the river can cut - ultimately of course, the sea. However, this is not a fixed entity. **Negative changes** to base level, whereby it falls or drops, may occur due to tectonic uplift of the land or isostatic recovery. Likewise sea level may fall as in **eustatic** changes during Ice Ages (see *Coasts*). **Positive changes**, whereby it rises, in contrast, reflect tectonic sinking of the land, or rising sea level due to post-glacial melt or global warming.

Negative changes increase the altitude of the source and thereby potential energy of the river. **Rejuvenation** results, with the river regrading itself back upstream from the coast. The **knickpoint** marks the maximum extent of the newly graded river at any one time. Beezley Falls on the River Greta near Ingleton in the Yorkshire Dales represents such a point, with paired river terraces of identical height marking the old flood plain. **Incised meanders**, such as along the River Lyn in Devon, and **gorges** such as the Avon Gorge in Bristol, often result from such negative base level changes, which if rapid enough may leave rivers cascading into the sea as waterfalls over recently heightened cliffs.

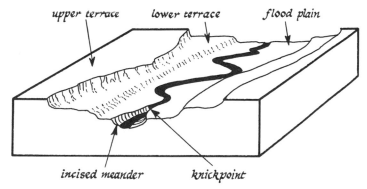

Positive changes have less dramatic effect on the long profile. The decreasing gradient simply encourages deposition and braiding. The Humber Estuary, for example, was formed by a rise in sea level leading to the wide, low-lying tracts of alluvial deposits from the rivers Trent, Ouse, and Aire in the Goole area.

6
GLACIATION

The study of glacial processes enables us to both understand landforms created in past climates and to comprehend aspects of present climatic change relating to the melting of glaciers or their growth. This may have important practical applications, such as in the planning of hydro-electric power (HEP) schemes. For example, at Grande Dixence in the Swiss Alps meltwater from twelve glaciers is fed via tunnels to a reservoir and dam for winter electricity generation.

SELECTED DEFINITIONS

Ice sheet (ice cap): Ice of continental proportions such as over Greenland and Antarctica today, the latter's ice being 1200-3000 m thick. Note, however, that the North Pole is floating sea ice.

Valley glacier: A river of ice occupying a pre-existing valley, such as the Rhine glacier.

Corrie glacier: Occupies a small armchair-shaped mountain basin and usually feeds a valley glacier.

Piedmont glacier: Valley glaciers spreading and merging to cover a lowland, such as the Malaspina Glacier in Alaska.

Whenever accumulation of snow exceeds melting, ice can build up. This happens above the **snow line** which delimits the lowest altitude of permanent snow, which does not melt in summer. This height will vary according to latitude. For example, at the poles it is at sea level. In the Alps it averages 2700 m - whilst in the Himalayas, 4500 m.

Glaciations

The most recent Ice Age, known as the **Pleistocene**, which started 2 million years ago, was the last of a succession of 'climatic accidents'. The present **interglacial** period will undoubtedly be followed by another Ice Age in near geological time, albeit likely to be delayed by contemporary global warming related to the greenhouse effect.

The Pleistocene had four distinctive glaciations (periods of cooling and warming) within it – although up to twenty-one have been suggested. Three were major – the fourth, the Loch Lomond Readvance, minor. At its maximum extent, ice covered Britain to a line north of what is now the rivers Severn and Thames, with tundra, and so periglacial, conditions to the south.

EVIDENCE OF CLIMATIC CHANGE

The build up of large ice sheets is a relatively slow process, yet climatic change can be much more rapid. The evidence for this comes from a range of sources:

Historical accounts can be extended back for several thousand years.

Pollen analysis and **dendrochronology** (involving the identification of climatic changes from annual growth rings in trees) can extend back to the most recent deglaciation.

Drill cores of ice caps and sea floor sediments, complemented by **oxygen isotope measurements**, give a record extending back over 100 000 years.

However, all this evidence suggests that no single explanation of climatic change is acceptable. For example, does the sun's energy output vary? Certainly, cycles of changing sunspot activity, over 11 and 22 years, have been identified. Also, receipt of solar radiation, varying due to the shape of the earth's orbit in space, its tilt, and wobbling axis – all as studied by M. Milankovitch – may, likewise, be relevant. Injections of volcanic dust by violent eruptions such as Krakatoa in 1883 and Mount St. Helens in 1980 have demonstrated short-lived effects of atmospheric cooling, and dramatic sunsets. Even tectonic plate movements may be influential – by changing the arrangements of oceans, hence ocean currents, and continental relief – therefore altering patterns of wind, temperatures, and precipitation.

Glacier formation and movement

If snow collecting in upland hollows fails to melt in summer it becomes **névé** or **firn**. This has a granular structure containing trapped air. Subsequent snowfall weighs the permeable névé down, squeezing the air out. This process over 20–30 years will form glacier ice – a crystalline, impermeable solid with special properties of flow.

Glaciers may be classified as **warm-based (temperate)** or **cold-based**. Both move downhill by gravity – faster and more erosive in the former due to flow or slipping, because the base is near enough to melting point to contain water. Cold-based glaciers (below the pressure melting point – because the weight of ice increases pressure and so lowers this critical temperature) move much slower, by internal flow, and cause less erosion. Both move faster at the surface and away from the valley sides, at rates ranging from mere centimetres, to 98 metres daily, recorded by Alaska's remarkable Bering Glacier.

Rhône glacier study to demonstrate the effect of sidewall friction

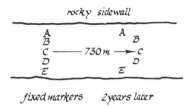

The above suggests that glacial movement involves many processes:

1. **Basal slip** represents sliding along the rock/ice interface and accounts for 50 per cent of forward movement – although up to 90 per cent in some cases. It is greatest when the bed is lubricated by meltwater in spring and summer and non-existent in cold-based glaciers which may be frozen permanently to the bedrock.

2. **Regelation** represents ice melting due to pressure on the upslope side of an obstruction and refreezing on the downslope side – just like weighted cheese-wire pressing through an ice cube.

3. **Fracture** and **faulting** lead to ice falls, usually where changes occur in the bedrock gradient or where the valley is constricted. The top 30 m of ice is normally affected.

4. **Flow** occurs in ice over 30 m thick by becoming **plastic** due to internal crystal deformation. This allows uphill movement for short stretches. There are two types of flow – extending and compressing:

(i) **Extending (accelerating) flow**, over a steepening, often convex bed, is associated mainly with the accumulation zone where erosion is less likely as debris moves downward. The top of ice falls is a common location.

(ii) **Compressing (decelerating) flow**, as gradients lessen, such as over a concave bed, is associated mainly with the ablation zone. This is common at the base of ice falls and at the glacier's snout.

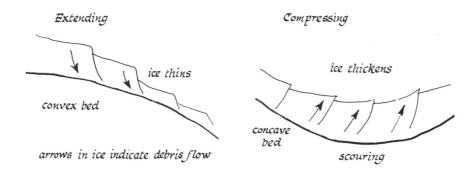

These processes of glacial movement become more evident and effective as the depth of ice increases and/or the gradient of slope steepens. Velocity changes are similarly logical in this respect – with faster flow on steeper slopes.

Advance and retreat

Advance – whereby the glacier's snout moves down the valley – occurs when accumulation exceeds ablation. (**Ablation** is all losses, principally melting, but also avalanche, calving, and evaporation.) It is notable that large glaciers are slower to respond to short-term climatic change than small ones. For example, glaciers in the Alps and Norway advanced from the late fifteenth to the late nineteenth century. Indeed, this **Little Ice Age** saw Britain very near to a major glaciation from the late seventeenth to the early nineteenth century!

Retreat – whereby the glacier's snout recedes up the valley – occurs when ablation exceeds accumulation. For example, glaciers retreated world-wide from 1967 to 1973 and evidence suggests contemporary retreat in the Alps.

Advance and retreat is never straightforward. A simple, steady advance or retreat, therefore, will rarely, if ever, be found. The advance and retreat of the Polar Ice Cap is particularly difficult to measure because the ice ends in 50 m high cliffs which **calve** into icebergs. However, whether this is due to lower temperatures causing advance, or warmer conditions encouraging rapid melting – and so calving – is not known.

Glacial budgets and systems

Glacial budgets refer to the balance between input (accumulation) and output (ablation). This will reflect whether the glacier is advancing or retreating. It is not the volume of the glacier, but its throughput. The budget is, therefore, a measure of activity - a positive budget demonstrating growth - a negative one, shrinkage.

The diagram below represents a budget in balance (equilibrium). The higher x is, the higher the budget - reflecting greater throughput.

A high budget glacier is a small but active system, transferring large volumes of ice. For example, Bassons in Chamonix (the French Alps).

A low budget glacier, by contrast, is not active. For example, polar glaciers where low temperatures are combined with low precipitation.

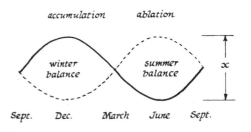

The **total budget** equals accumulation plus ablation.

Glaciers should be regarded as **open systems** of inputs, storage, and outputs. Any change in input will, therefore, be countered by a change in output.

Inputs (mainly to accumulation zone)		**Storage**	**Outputs** (mainly to ablation zone)
1. Materials:	Precipitation Avalanche Rock debris	Glacier consisting of ice crystals, debris, water, and air	Water Water vapour Ice (icebergs) Rock debris
2. Energy:	Gravity Solar radiation (insolation) and conduction Geothermal		Heat from friction

Processes of glacial erosion

Almost all the glacial processes of erosion are mechanical (physical) due to the cold temperatures limiting chemical reactions. Pre-glacial weathering is very important, allowing the ice to move already loosened material. Above the level of the glacier, frost shattering produces loose debris which may fall onto the ice and become part of its **moraine**. For example, **lateral** if occurring along the edges - **englacial** if fallen down crevasses or covered by subsequent snowfall (see later).

1. **Plucking (quarrying)** is caused by ice freezing round protruding rocks which are plucked away as the ice moves on. It is particularly effective on well-jointed rocks and in areas previously weathered, such as the backwall of a corrie.

2. **Abrasion (corrasion)** is the 'sandpapering effect' of angular material embedded in the glacier's sides and base. It needs a continuous supply of debris, from glacier to base, and produces smooth landforms scratched by parallel **striations** which indicate direction of ice flow by the groove fading out as the offending rocks wear away.

3. **Attrition**, so relevant to the formation of striations, is most evident as the load is ground down to **rock flour**. Rock flour carried in suspension gives glacial meltwater its ethereal milky appearance.

4. **Pressure release (dilatation** or **sheeting)** occurs after the weight of ice has been removed on melting. For example, recession of the Allalin Glacier at Stausse Mattmark in Switzerland illustrates this well. However, it is more effective during glaciation as rock is replaced by ice which is of lower density.

5. **Subglacial stream action** is not, strictly speaking, a glacial process, but removes material during spring and summer melting.

Once glacial erosion begins, there is **positive feedback**, with cumulative causation in consequence. For example, a small nivation hollow leads to a small glacier which erodes a bigger hollow - hence allowing more accumulation of snow, a bigger glacier, and more erosion! Corries (explained later), consequently, grow slowly at first, then much more rapidly. Likewise, valleys are at first deepened a little, then overdeepened. **Overdeepening** is erosion below base level - the sill in the case of corries, sea level in fjords. Compressive flow is particularly influential, especially when two or more tributary glaciers join and deepen the main valley. Basins and rock steps in the main valley floor are exaggerated by alternating compressive and extended flow, over basins and steps respectively, with the basins often overdeepened.

Some glaciers are far more erosive than others. In general, glaciers with a large total budget of accumulation and ablation are the most erosive, such as warm-based glaciers in the Alps. Glaciers with a small total budget tend to be the least erosive and may even protect the underlying rock. Cold-based glaciers in Greenland, for example, have been studied in this context. Other factors influencing rates of erosion include the nature of the bedrock, the original valley's shape, size, and gradient, the amount of pre-glacial weathering, and the quantity of moraine.

Features on the glacier's surface

Crevasses are longitudinal and transverse cracks produced when ice is subjected to marked tension. The latter are particularly common, resulting from extended flow when the gradient increases and the ice speeds up – so becoming thinner. Crevasses also occur as a result of fracturing and faulting in an ice fall.

Pressure ridges are ridges of ice pushed up by compression when the gradient decreases, causing the ice to slow down and thicken.

Seracs are blocks of fractured ice, usually at an ice fall, or where sets of crevasses and pressure ridges intersect.

Bergschrunds are the large cracks in the ice along the backwall, and occasionally sidewall, of corries (see later) which open in the summer and close in the winter. They allow entry of meltwater to the glacier's base.

Glacier tables are notably large, perched blocks of rock on the glacier's surface. Indeed, small rock fragments, if dark enough, may have the opposite effect by absorbing solar radiation, warming and so embedding into the melting ice.

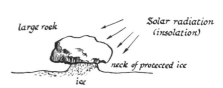

Moulins are 'swallow holes' on the glacier, cut by meltwater streams. They are thought to originate by a swirling action, in the same manner as river bed pot holing.

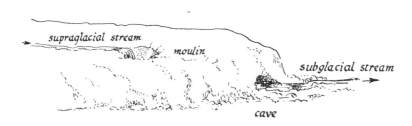

Moraine is any glacial debris ranging in size from large angular boulders to fine rock flour. It may be on the ice surface (**supraglacial**), within it (**englacial**), at the base (**subglacial**), or – if marking the maximum extent of snout advance in a transverse mound – **terminal**.

Ogives are curved, corrugated bands of alternating shade. The curved bands of dark colour are troughs up to 1 m deep and represent summer ice containing more debris. The alternating clear white bands, however, stand up as ridges – their higher albedo causing less melting – and represent winter accumulation.

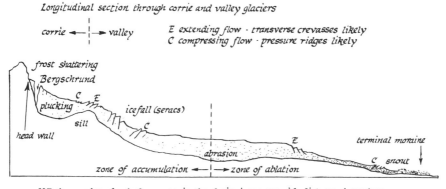

Longitudinal section through corrie and valley glaciers

corrie ←─┊─→ valley   E extending flow · transverse crevasses likely
                       C compressing flow · pressure ridges likely

N.B. the number of rock fragments in the glacier increases with distance downslope

Features of erosion in uplands

Upland erosion is best exemplified in areas of resistant rocks such as in the Alps and the Lake District. These harder rock areas were of high altitude before the Pleistocene and so became source areas for glaciers. Indeed, such is the resistance of the rocks that they have preserved the features of erosion well.

Pre-glacial relief and drainage patterns were particularly important of course because the ice became channelled down existing valleys. Therefore, comprehension of glacial features can only be complete when one also knows the pre-glacial form and understands the process of change.

Corries (cirques or **cwms)** start as sheltered, gently sloping hollows on the shady side of a mountain where snow accumulates. It is no coincidence, for example, that corries of a north to north-east orientation predominate in the Lake District and are found at lower altitudes than those of warmer aspect facing the south-westerly prevailing winds. The snow-névé-ice sequence discussed earlier is followed by all the processes listed below happening together:

♦ The headwall is weathered and frost shattered, generating rock fragments which fall onto the ice and down the bergschrund, so providing the 'tools' for abrasion.

♦ Seasonal and diurnal melting allows water to flow down the bergschrund which by alternate freezing and thawing causes disintegration of rocks at the bottom.

♦ Ice starts to move downhill under gravity, plucking loose rocks from the base and abrading solid rock with the grinding action of the tools referred to earlier.

♦ Summer meltwater helps the ice to move by lubricating the base.

♦ The volume of ice increases and pressure release sheeting occurs on the backwall and base. Blocks are plucked from well-jointed rocks and the corrie is overdeepened.

♦ Under pressure from the volume of ice, and helped by lubricating meltwater, the corrie glacier rotates enough to move uphill and over the **rock lip** or **sill**. This rotational movement involves compressive flow (C) with some fracturing.

Corrie in section

back wall   c   *the dots represent bands of dirt which*
                   sill   *are tilted progressively to give*
                              evidence of rotation

After glaciation a deep armchair-shaped hollow is left cut in the mountainside which may be occupied by a small lake - a **corrie lake** or **tarn**, such as Red Tarn on Helvellyn in the Lake District.

Arêtes start as a ridge between two hollows on a mountainside. As the hollows are deepened to form corries, the ridge gets steeper. Above the ice, and so exposed to frost shattering, it becomes sharper until standing as a prominent 'knife edge' which will remain after the glaciation. For example, Striding Edge above Red Tarn is only now being blunted by thousands of heavily booted walkers!

Horns start as high altitude ground - often a nunatak above the ice. (**Nunataks** are mountain summits left protruding above the ice.) Steepening of corrie headwalls on all sides, although usually three, and frost shattering on the summit, leaves a pyramid-shaped peak in the centre. This horn remains prominent after the glaciation. Mount Snowdon in Snowdonia is a good example of a nunatak becoming a horn. However, the Matterhorn overlooking Zermatt in the Swiss Alps is a classic.

NB: Frost-shattered peaks generally remained above the level of the ice as nunataks. However, some summits were covered by ice and became smoothed and rounded. Since these did not experience frost shattering, they were in some senses protected by the ice. The Cairngorms in Scotland and the Mourne Mountains in Northern Ireland are good examples.

Troughs start with a pre-glacial V-shaped river valley which is deepened, and its sides steepened, by the erosive action of the glacier into the characteristic U-shaped glaciated valley or trough.

At the top of the steep section a
break of slope marks the level of
the ice, above which the **benches**
or **shoulders** flatten out.

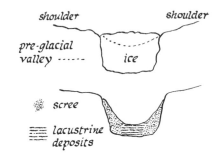

After the ice has melted troughs
may be modified to give a cross-
section more like this:

Trough ends start where several corries converge and supply ice to the
same valley outlet. The combined weight of this ice has greatly increased
erosive power, so overdeepening the main valley and creating a sudden
increase in gradient down to the floor of the trough. This sharp break of
slope is later softened by weathering - all illustrated by Grisedale trough,
supplied with ice from two corries, now known as Nethermost Cove and
Grisedale Tarn, again in the Lake District.

Truncated spurs before glaciation were the protruding interlocking
spurs of a river valley. However, ice is far more powerful than water and
much less flexible. Consequently, a glacier following the valley would
truncate or 'behead' the spurs leaving a line of steep cliffs which may
remain evident today along the sides of a glacial trough. The southern
slopes of Blencathra in the Lake District show this well.

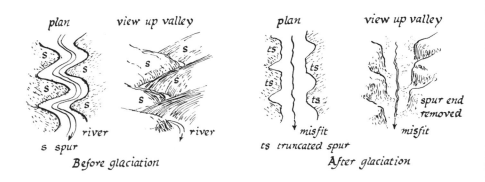

NB: After glaciation the river is very small relative to the size of the new valley. Since it
has insufficient energy to have eroded such a large valley it is called a **misfit (underfit)**
river. Grisedale Beck, for example, is a small stream occupying a large glacial trough.

Hanging valleys relate to pre-glacial tributary streams feeding the main river channel. Glacial overdeepening of this main valley leaves the less eroded tributary valley high on the shoulder. After melting, this tributary valley is left hanging above the main trough, its stream or river cascading as a waterfall over the edge. Whelpside Gill and Comb Gill hanging above Thirlmere in the Lake District are good examples.

Ribbon lakes need a glacial trough to be blocked at one end by, for example, moraine, in order to create a natural dam which allows a long, thin lake, such as Thirlmere, to form behind it. Most today are being filled in by progressive sedimentation as weathered debris is removed from surrounding slopes. Indeed, Buttermere and Crummock Water in the Lake District have been formed by the division of one lake, by deposition in the centre.

Roches moutonnées start as a large block of resistant rock in the path of a glacier. The ice passes over and around the rock forming a smoothed stoss side (upstream) with striations caused by abrasion. The lee side (downstream) is left rough and steeper by plucking. Usually unvegetated, they are still visible today albeit modified by weathering. Both Tourbillon Castle and the fortified Church of Valère, prominent landmarks of Sion in Switzerland's Rhône Valley, are built on such features.

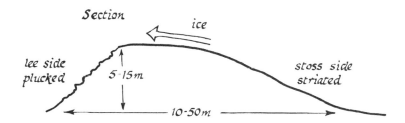

Features of erosion in lowlands

Ice scoured basins (lochans) are often now filled by lakes, such as Loch Avon in the Scottish Cairngorms. Pre-glacial drainage patterns were obliterated by the ice leaving irregularly shaped hollows and a deranged pattern. Weathered rock was stripped away leaving bare rock and thin soils. The resulting **knock and lochan topography** (the 'knock' being rocky hillocks) is well illustrated across the Canadian and Baltic shields. Similarly, well-jointed Lewisian gneiss in the Scottish Hebrides produces a regular pattern of knock and lochan hollows and drainage.

Crag and tail features could be argued as depositional in origin, but erosion seems to be the key to their explanation. Highly resistant rock obstructions, such as volcanic plugs, protect the softer tail on the lee side from erosion.

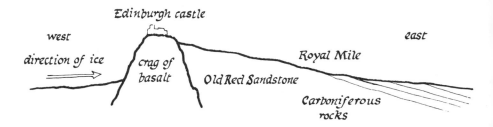

Fjords are normally explained as being *'drowned glaciated valleys'* caused by the post-glacial eustatic rise in sea level flooding a glacial trough. However, this does not explain why many fjords (and indeed many lakes) have their beds overdeepened by erosion to below sea level. Examples in Scotland, Norway, Chile, and New Zealand all show excessive overdeepening and a rock barrier (**threshold**) of shallow water at the mouth. However, before suggesting possible explanations, it would be appropriate to remember that the discrepancy and overdeepening outlined above will not be as great as might appear because sea levels are always lower during Ice Ages.

A combination of factors probably causes the overdeepening, as shown in the cross-sections which follow:

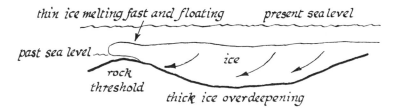

The glacier, melting quickly on meeting the sea, would consequently be thinner and so incapable of overdeepening. Indeed, it would be likely to float at the snout causing no erosion at all. Hence thick ice inland overdeepens the bed and then flows uphill to the snout at the threshold.

Pre-glacial freezing of saturated ground may also have some relevance, as suggested by the next cross-section. The freezing and weakening of the saturated layer makes it particularly vulnerable to glacial erosion - hence its removal with resultant overdeepening.

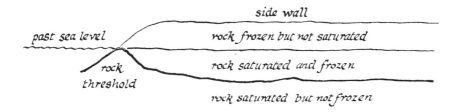

Some rock thresholds could simply be terminal moraines, hence even these are the subject of debate.

NB: Care must be taken over distinguishing between fjord and fjard. **Fjard** is a term used to describe drowned glaciated lowland river systems such as Strangford Lough in Northern Ireland.

Features of deposition

Glacial deposits are removed from uplands, whereas they remain in lowlands. Hence the association of deposition with lowland areas. The collective term **drift** covers both **direct deposits** - till from ice, and **indirect deposits** - fluvioglacial material from meltwater.

Direct deposits (till) were deposited by ice, as opposed to water, and are characteristically poorly sorted, unstratified, angular, and compacted.

◆ **Lodgement till** is the glacier's base load 'plastered' onto the ground because its weight has become too great. Frequently called **boulder clay**, although it does not necessarily consist of clay with boulders in, it is normally deposited in sheets consisting of a matrix of fine rock flour containing angular stones of all sizes - including boulders. Most till sheets are formed from lodgement, but some are from ablation.

◆ **Ablation (melt out) till** is a combination of subglacial, englacial, and supraglacial moraine released by a stagnant glacier as it decays on melting - losing its volume by vertical shrinkage. Although there is some evaporation, most water percolates down before escaping, leaving the debris unsorted - effectively dumped *in situ*.

Till sheets vary in thickness and generally mask all existing relief. For example, over Holderness in England the till is 60 m thick! Analysis by pebble orientation studies may reveal the source of the ice sheet or glacier because the long axis of each pebble is usually parallel to the direction of ice movement and angled down towards the origin.

Till fabric analysis revealing imbricate structure (like roofing tiles)

In the field, therefore, a compass to measure orientation (later to be recorded on a star or rose diagram) and a clinometer to measure angle of dip may reveal much about the till studied. Cliffs along the Holderness coast, for example, allow easy access and show that some till originated in Norway, whilst other samples came from the Lake District. Analysis of the stones, which include Norwegian laurvikite and Lake District Shap granite, adds further evidence.

◆ **Drumlins** are smooth, oval hills of till found in **swarms**. Although of debatable origin they are likely to have formed like ripples under deep, mobile ice. Indeed, their degree of elongation is related to the speed of ice movement – the direction of which can be determined by their orientation and till fabric analysis.

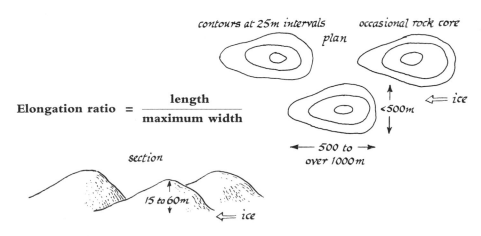

◆ **Moraines** are landforms made of till found both in glaciated lowlands and in upland valleys once occupied by ice. Variable in height, up to 50 m, they are complicated by a number of advances and retreats made by the ice that transported the debris.

(i) **Terminal (end) moraines**, such as the one blocking the outlet of Malham Tarn, mark the maximum advance of the ice sheet or glacier.

   (a) **Recessional (stadial) moraines** are left where the ice front stood still for a time before receding further. Consequently, a succession of recessional moraines, usually parallel to the terminal moraine, mark stages in the retreat of the ice front.

   (b) **Push moraines** are terminal moraines that have been 'bulldozed' forward by an advancing ice sheet or glacier. The compressive flow of snout ice gives this debris a marked imbricate structure.

(ii) **Medial moraines** are occasionally found in the centre of a valley. They are longitudinal moraines running down the centre of a glacier, having formed by the merging of lateral moraines from two tributary glaciers joining together. However, they are very vulnerable when the glacier melts and so generally short-lived.

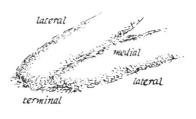

(iii) **Lateral moraines** are much longer lasting – effectively embankments of frost-shattered debris running along the valley sidewalls.

(iv) **Rogen (washboard) moraines** are transverse ridges across the glaciated valley likely to be of subglacial origin. Up to 30 m high, they may be very difficult to distinguish from recessional moraines. However, small ones, 2-5 m high, may be related to ogive bands.

◆ **Erratics** are large boulders, 1 m or more across, carried by ice and deposited far from their origin. Analysis of the erratic may allow the source area to be determined and so clarify the direction of ice movement. For example, Carboniferous limestone erratics from Fountains Fell occur on shales near Malham Tarn.

Indirect deposits (fluvioglacial) were deposited by meltwater and are characteristically well sorted, stratified, rounded, and loosely compacted.

All meltwater streams have heavy loads and generally slight gradients – conditions suitable for abundant deposition. Nearest the ice front the coarsest deposits are dumped first with progressively finer material deposited further and further away.

◆ **Eskers** are long sinuous ridges of sand and gravel, rather like railway embankments, up to 30 m high, and at 90 degrees to the ice front. They are never beyond the ice limit because they formed in tunnels under the glacier. They are, therefore, the deposits of subglacial meltwater streams flowing as the glacier melted.
Indeed, they are revealed progressively as the ice 'melts back' and may be beaded, with each bulge or bead representing a halt in the ice recession.

Plan

◆ **Kames** are irregular delta-like mounds of sand and gravel occurring along the melting ice margin. They can also be formed where the load of a supraglacial channel or pool is lowered to the surface as the ice melts around it.

◆ **Kame terraces** are similar, but in ridges along the sides of valleys. They have been deposited by meltwater streams flowing between the melting glacier and the valley wall. Unlike lateral moraines, but as with kames, these deposits are sorted.

stream sequence

ice

Section during deglaciation

Kame terraces

Section today

◆ **Outwash deposits** are horizontally bedded sheets of sediments ranging from fine rock flour to gravel. They are always downward of the ice front.

(i) **Outwash plain (sandar)** is a flat but deep sheet of sediment deposited by summer melt or deglaciation. Braided streams normally cross the plain. Relict sandar is found near Blakeney in North Norfolk.

(ii) **Valley trains** are similar sediments, but confined to a valley and often dissected and terraced by later river action. Good examples are found in the Chiltern Valleys north of the River Thames.

◆ **Kettle holes**, like eskers, are never found beyond the ice front. They were formed when large blocks of ice left behind by a receding glacier, and often partially buried by fluvioglacial meltwater deposits, eventually melted – leaving depressions likely to fill with water. The lakes in Rise Park on Holderness are a good example.

◆ **Varves** are rather special and very revealing deposits on lake beds. The sediments were deposited by meltwater entering a **proglacial lake**, found near the ice margin. Rapid summer melting allows thick layers of coarse sand to build up. As meltwater discharge decreases to virtually non-existent in winter, a thin layer of fine silt settles on top. Each band of thick coarse and thin fine material represents one year – a varve, counting and analysis of which may reveal the date of the lake's origin and variations in the rates of annual melting. Freshly-exposed varves may be studied in the marine cliffs at Barmston on the Holderness Coast.

◆ **Meres** are lakes formed in hollows on the sheets of outwash or till beyond the ice front. Hornsea Mere on Holderness, for example, was originally three times its present size.

Periglacial processes and landforms

The definition of **periglacial** is rather vague, literally meaning '*round about the ice sheet.*' It is best applied to areas, past and present, with a tundra climate and permanently frozen subsoil (**permafrost**).

Processes

1. **Weathering**: Frost shattering depends upon the lithology and degree of jointing to create distinctive scree (talus) deposits. It is most effective with widely fluctuating temperatures, especially a large diurnal range, at high altitudes.

2. **Mass movement**: Solifluction under gravity can act upon slopes as low as 2 degrees because the waterlogged summer surface slides easily over permafrost and vegetation is often absent, or at least so shallow-rooted as to be useless as an anchor.

NB: Some authors prefer to use a term more specific to periglacial climates, namely **gelifluction,** leaving solifluction as the more general 'soil flow'. Also, one finds alternative spellings - gelifluxion and solifluxion. All the terms are, in effect, synonymous.

3. **Removal**: Saltation and deflation (by wind) may blanket existing relief. The sand deposits (by saltation) over Breckland, East Anglia, and the fertile, unbedded, and stone-free loess (by deflation) over the Bonn and Cologne area of Germany are excellent examples.

Landforms

Thermokarst features are irregular mounds and hollows caused by the formation and melting of ground ice. Poor drainage, with many marshes and lakes, results.

Pingos are isolated dome-shaped hills interrupting characteristically flat tundra plains. Recent research suggests different modes of formation, with the so-called **open system** explaining pingos in areas of continuous permafrost, such as East Greenland, allowing frozen surface lakes. **Closed system** explanations state that no surface water is involved and refer to areas of discontinuous permafrost, such as the Mackenzie Delta of northern Canada. Whatever their precise origin, the following cross-sections demonstrate the nature and scale of these landforms.

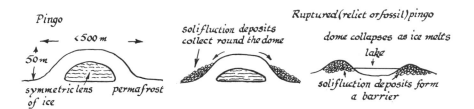

Ice wedges form in unconsolidated material which has frozen and expanded in winter only to contract and crack on thawing in summer. Water subsequently filling the cracks will freeze and expand next winter as ice wedges, which may grow over the years to enormous proportions. **Fossil wedges** of sand and silt, fallen and washed by meltwater into the thawed wedge hollows, may reflect past periglacial conditions. Good examples are exposed in a cutting of the M40 through the Chilterns in southern England.

Polygonal stone patterns and **stone stripes (garlands)** also form by alternate freezing and thawing. The centres of the polygons are domed during expansion, due to **frost heave**, allowing stones and other debris to move through gravity to the sides. The steeper the slope angle, the more elongated the polygons – ultimately leading to garlands.

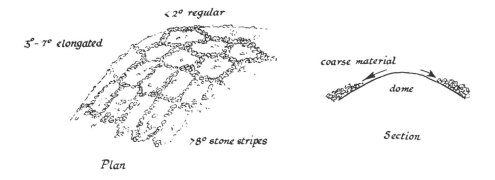

Lakes may be trapped by landslip debris.

THE IMPORTANCE OF PERIGLACIAL PROCESSES

Periglacial processes need to be understood if practical solutions are to be found for the building and engineering problems associated with contemporary periglacial conditions. For example, infrastructure, settlement, and oil pipeline development in Alaska (involving stilts, gravel pads, and fibrous insulation amongst other innovations) has been developed with increasing success as periglacial research has progressed.

Interpretation of present landforms in humid temperate climates, such as in Britain, is only possible if we understand periglacial processes because there are so many 'fossil features' from past, colder climates to comprehend. Solifluction features in North Yorkshire, for example, are mostly relicts from periglaciation associated with rotational landslips. Likewise, solifluction deposits (known as **head**) fill valleys in the Yorkshire Wolds. Similarly, dry valleys, such as the Watlowes above Malham Cove, cut during conditions of permafrost, can only be appreciated with periglacial understanding.

Finally, it is worth noting that the unsorted, angular debris associated with glacial and periglacial deposition can only be differentiated if one remembers that periglacial deposits are of looser texture, and, because they rarely move more than 1 or 2 km from their source, will not contain erratics.

7

COASTS

SELECTED DEFINITIONS

Coast: The interface of land and sea. It is the narrow zone where land and sea overlap and intersect. It includes cliffs, sand dunes, and the shore.

Coastline: The plan view of the coast showing the degree of indentation.

Shore: (a) **backshore** - usually beyond wave influence but invaded during spring tides.

         (b) **foreshore** - inter-tidal zone between high and low water.

         (c) **nearshore** - breaker zone below low water mark.

         (d) **offshore** - beyond wave influence.

Beach: Any loose material accumulated on the shore.

Littoral: Meaning 'of the shore'.

The coastal landscape is dependent on:

- **lithology**
- **structure** - dips, faults, and so on
- **processes** - wave action, sub-aerial weathering, coral growth, volcanic activity
- **climate** - past or present glaciation causing changes in sea level
- **human activities**

The coast may be considered as an equilibrium system of inputs, storage, and outputs.

Inputs	Storage	Outputs
1. Sediments from rivers, cliffs, and sea bed	The beach (Note that equilibrium is maintained by changes in beach form)	Sediment loss by longshore drift or aeolian processes
2. Energy from wind (waves) and tides		

Waves

Waves are generated by wind blowing over the sea. Their height, steepness, and period are of particular significance in determining coastal geomorphology.

Wave height depends on the wind's strength, duration, and **fetch** (length of open water). Where the fetch is long, such as over the Atlantic Ocean, storm duration will determine height. If the fetch is short, however, such as over the English Channel, storm duration will only increase the wave height to a limited extent.

Wave steepness is more influential than height alone.

$$\textbf{Wave steepness} = \frac{\textbf{height}}{\textbf{length}}$$

Wave period is the time elapsed between one wave breaking at a given point and the next.

Waves are energy systems. In open sea the water particles are in orbit, as **waves of oscillation**. As the waves approach the shore, however, the orbits become elliptical through bed friction in the shallowing water. The water particles consequently move forward, as **waves of translation**, until the unsupported crest breaks.

Breaking waves may be thought of as 'tripping themselves up' on the shelving nearshore sea bed.

Prevalent waves are those of greatest frequency. In Britain they reflect the prevailing south-westerly winds.

Dominant waves, however, are the strongest and so therefore have the greatest effect. In Britain these are north-easterly, affecting the east coast.

Storm surges occur in the North Sea when a high tide coincides with a northerly wind because the English Channel acts as a bottle neck causing the water to pile up. Early in 1953, for example, a deep depression causing gale force winds coincided with a spring tide. The resulting storm surge had disastrous consequences for both the Netherlands and East Anglia where coastal defences proved inadequate to prevent extensive flooding with appalling loss of life and property.

Wave refraction concentrates wave energy on headlands, causing erosion, whilst dissipating it in the bays, allowing deposition.

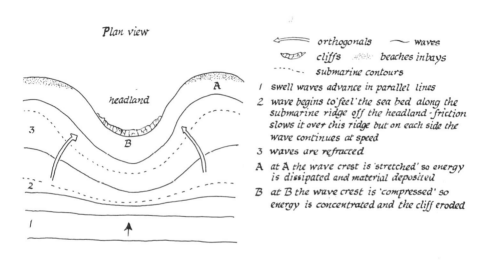

Plan view

← orthogonals ∼ waves
cliffs beaches in bays
submarine contours

1 swell waves advance in parallel lines
2 wave begins to 'feel' the sea bed along the submarine ridge off the headland - friction slows it over this ridge but on each side the wave continues at speed
3 waves are refracted
A at A the wave crest is 'stretched' so energy is dissipated and material deposited
B at B the wave crest is 'compressed' so energy is concentrated and the cliff eroded

Constructive (surging or **spilling) waves** are allowed to run their course without interference from those behind. Flat and low in height, their **swash** carries material up the beach, forming a **berm**, with the energy consequently dissipated over a wide area - hence the weak **backwash**.

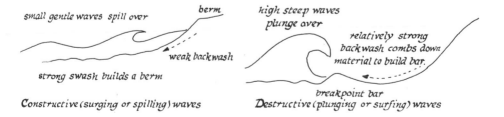

small gentle waves spill over   berm
weak backwash
strong swash builds a berm

high steep waves plunge over
relatively strong backwash combs down material to build bar.
breakpoint bar

Constructive (surging or spilling) waves   *Destructive (plunging or surfing) waves*

Destructive (plunging or **surfing) waves** break more frequently, with the force of the swash reduced by the previous wave's backwash. These steep waves plunge onto a smaller area, so concentrating their energy. Their backwash prevails, given little time for infiltration, and will carry shingle down with it to form a **breakpoint (longshore) bar**. Note, however, that some material may be flung above the high tide mark forming a **storm beach** on the backshore.

Beach material

Constructive waves are associated with the progressive steepening of a gently sloping beach. Destructive waves, by contrast, lessen its profile by a combing down process. However, it is not solely the nature of the waves that determine beach form – sediment size is more influential in that the coarser the material, the steeper the profile.

Shingle, for example, can retain a relatively high slope angle and allows greater infiltration. Swash, therefore, exceeds backwash and carries material upwards.

Sand only retains low slope angles and has lower infiltration. Consequently, most of the swash returns as surface backwash – carrying sand with it.

Silt (mud) makes the flattest beaches.

Processes of erosion

1. **Hydraulic action** is associated with **wave pounding**. Air trapped and compressed into joints by the encroaching water expands with explosive force when the wave recedes. Likewise, high pressure spray may be forced into cracks – hence weakening the joints, fault lines, and bedding planes of sedimentary rocks, such as the chalk at Flamborough Head.

2. **Abrasion (corrasion)** is the most effective method of coastal erosion with cliffs being worn away by pebbles, shingle, and finer material being hurled against them.

3. **Attrition** of coastal boulders, pebbles, and shingle breaks them down further, by impact and friction associated with rolling, into smaller, finer fragments.

4. **Weathering** operates in the inter-tidal zone and up to the level of flying spray:

(a) Mechanical – salt crystallisation enlarges cracks and pores subject to frequent wetting and drying.

(b) Chemical – solution, caused by the corrosive (solvent) action of sea water, plus the acids produced by algae and guano, is most effective on calcareous rocks.

5. **Sub-aerial processes** of mass movement and frost shattering on resistant cliffs – soil creep, slumps, and slides on softer rocks add to the processes of erosion operating along the coast. These processes may be exacerbated by human activities such as building on cliffs or removing beach material – consequently disturbing the coastal equilibrium.

Landforms of erosion

Cliffs will be undercut at the base to form a **wave cut notch** leading to collapse by fall or slump depending on the lithology and dip of predominantly resistant rocks. Progressive retreat of the cliff will leave a gently sloping inter-tidal **wave cut (shore) platform** scarred by abrasion and solution. The wider this feature becomes, the more wave energy it will dissipate, so limiting erosion – just as accumulation of cliff debris does.

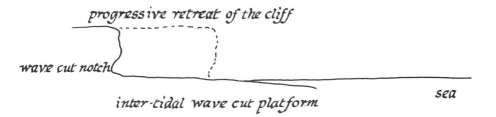

Caves will develop where lines of weakness such as joints or faults may be exploited.

Blowholes start with partial collapse of the cave roof following enlarged joint expansion to the cliff top aggravated by hydraulic action.

Geos result if the whole cave roof collapses leaving a long narrow inlet.

Arches form on headlands where caves cut back to back.

Stacks result when the arch collapses.

Stumps remain once the stack collapses.

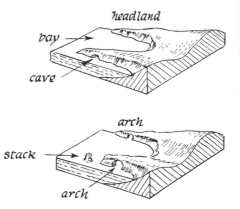

All these features are demonstrated memorably in the particularly resistant chalk of Flamborough Head - the **geo** north of Selwicks Bay, Adam (**stack**) and Eve (**stump**) on the extensive **wave cut platform** nearby, **wave cut notches**, and the **arches** through Greenstacks - not least Pigeon **blowhole** with its crater of slumping, overlying boulder clay.

Poorly consolidated rocks, such as the boulder clay of the Holderness coast south of Flamborough, erode rapidly. Indeed, this is Europe's fastest eroding coastline. Both the cutting of the wave cut notch and the resulting cliff fall are very rapid with rotational and non-rotational slides dominant in the clay lubricated by heavy rain. Sometimes the clay liquefies, with mudflows descending to the narrow beach where strong waves and longshore drift allow little accumulation to protect the cliffs. Consequently, 3-5 km of Holderness has been lost since Roman times. Most is lost in storms with an average recession of 2 m per year - hence the loss of nearly 50 villages recorded in the *Domesday Survey* of 1086.

Processes of transport

As with slopes, if no transport removes the eroded material it will accumulate and reduce further erosion by acting as a buffer - an illustration of negative feedback. Transport either moves material up and down the beach or along the coast.

Sediment size and the nature of the waves determines shore profiles (which are a section from offshore to backshore). Sediment variations, for example, are associated with characteristic, if often ephemeral, landforms:

Berms are usually flat-topped ridges of shingle marking the limit of the high tide swash.

Breakpoint (longshore) bars contain fine sediment which collects just into the nearshore zone where the water is shallow enough to interrupt the approaching waves' oscillations and cause them to break.

Storm beaches form inland of the berm. They comprise larger pebbles and cobbles hurled above normal high tide level, and stranded, by storm waves.

Cusps are rhythmic beach features composed of coarse ridges and fine runnels. Their origin is rather obscure, but once formed it is easy to see why they are maintained. The swash divides at the ridges and runs off to return as backwash scouring the runnels.

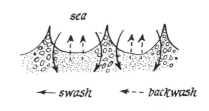

Shore profiles can change rapidly depending upon the prevailing conditions. Each beach constantly changes, therefore, to reflect the shore's aim of achieving equilibrium.

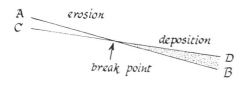

In winter destructive storm waves may flatten the original profile AB to CD.

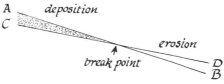

In summer's calm weather the new profile CD may be steepened again to AB.

Clearly the response of a beach to changing wave conditions is impossible to observe during fieldwork. However, they can be simulated in a laboratory wave tank.

BEACH PROFILE CHANGES IN THE WAVE TANK

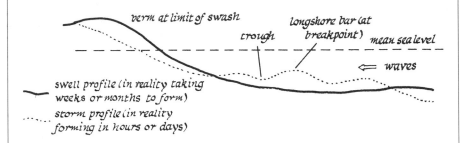

1. The tank was set up with sediment of uniform size (fine sand), arranged to form a profile mid-way between the two shown above.

2. Low flat waves were then produced in order to simulate a swell.

3. Sand removed from below the break point was transported to build a berm in the upper part of the beach.

4. The equilibrium **swell profile** was reached resulting in no further change.

5. The wave height was then left constant, but the period decreased in order to allow the waves to arrive in quicker succession.

6. The upper beach and berm then progressively eroded whilst a breakpoint bar built up, with a trough on the shore side of it.

7. Eventually the equilibrium **storm profile** was reached leading to no further change.

Clearly there are simplifying conditions in this demonstration - not least the uniform sediment size and no tidal variation or longshore drift. However, the concept of **dynamic equilibrium** is certainly well illustrated. In reality, steep waves, with high energy input, create a flat, wide beach beyond a breakpoint bar which causes the waves to break early - with the extensive beach beyond absorbing the energy, allowing equilibrium to be reached.

Transport along the coast effectively moves material along the shore:

Beach drifting operates mainly in the surf zone, and is caused by an oblique approach of waves and, consequently, swash, followed by direct return to the sea of the back-wash. Sand, shingle, pebbles, and so on are moved progressively along the beach.

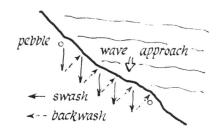

Longshore currents develop below the breakpoint in the general direction of wave travel.

NB: The term **longshore drift** should be applied to the combined effects of beach drifting and longshore currents.

Due to its funnel shape and northerly approach of the tides, longshore drift in the North Sea is southwards. On the English Channel coast it is eastwards, and on the Norfolk coast, westwards towards the Wash. However, reversals of longshore drift direction may occur for long enough periods to cause contradictory evidence in the analysis of beach material origins.

Tidal streams are caused by local oddities in coastal topography causing differences in the tide level. For example, a strong tidal stream runs between two sides of the Lleyn peninsula between Bardsey Island and the mainland.

Rip currents are very strong and caused when longshore currents are channelled by sea bed relief, say on meeting a headland, to turn out to-wards the sea. Consequently, they are particularly dangerous to bathers.

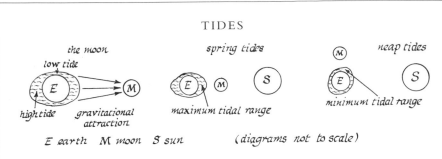

TIDES

the moon   spring tides   neap tides

low tide

high tide   gravitational   maximum tidal range   minimum tidal range
             attraction

E earth   M moon   S sun   (diagrams not to scale)

Both the sun and the moon, the latter to a greater extent, exert a gravitational pull
on the rotating earth's surface. This causes a rising and falling motion in the waters
of the larger oceans, so producing tides. The difference between high and low tide
is known as the **tidal range** and this will vary from a maximum **spring tide** when
the sun and moon are in a straight line, and so pull together, to a minimum **neap
tide** when the sun and moon are pulling at right angles. The greater influence of
the moon dictates approximately two high tides a day, with spring tides twice each
lunar month - therefore once every fourteen days.

Control of erosion and transport

Sea walls provide direct protection from wave pounding, but will require
replacing within 25-150 years because they act like a cliffed headland and
concentrate wave energy - so destroying themselves. It is notable that such
traditional 'hard' defensive strategies of solid concrete walling are adopted
less often nowadays. This is due to their high cost and the greater
effectiveness of energy dissipating **rock armour** - such as the massive
blocks of Swedish granite protecting the village of Mappleton, south of
Hornsea on the Holderness coast.

Groynes hold beach deposits
in place and so protect the coast.
However, the downside is starved
of material and erosion is aggravated
further down the coast.

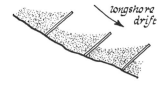

NB: Aggravated erosion beyond defended stretches of coastline is a major concern given
the financial (and political) cost of coastal protection decisions. For example, within
months of completing the Mappleton defences, severe erosion was measured to the
immediate south.

Anchor measures such as sandbags, iron spikes, and even plastic seaweed have been tried with varying degrees of success along the Holderness coast. Likewise, experimental **cliff top drainage**, to limit slumping, has been evaluated - but this does not stop the loss of beach material. The use of energy-dissipating **offshore reefs** of mine waste, such as coal slurry, is now being suggested as another cost-effective alternative to traditional engineering solutions. However, the only proven example of 'soft' engineering, and certainly more effective than groynes, is **beach nourishment**. This is the artificial input of material, such as sand, to compensate for natural losses. Miami, Florida, is nourished in this way, but it is normally prohibitively expensive.

Processes of deposition

Sediment for deposition comes from three main sources - river, cliff, and sea bed erosion. It is stored on or near the shore as beaches, spits, dunes, and so on. Of note today is that most coastal sediments are suffering a net loss - back to the sea bed from where they probably came at the end of the Pleistocene Ice Age when sea levels rose rapidly.

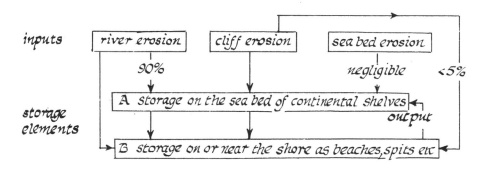

Of the three inputs, river erosion supplies the vast majority - most of which goes into A. Cliff erosion supplies less than 5 per cent, mostly into B, whilst sea bed erosion is negligible today, compared to the post-glacial period referred to above.

Output from the shore to the sea bed occurs during storms by destructive waves, rip currents, flow along submarine canyons, and by longshore drifting reaching deep water, such as at the end of a spit.

The features of deposition on or near the shore can only be created today by constructive waves. Indeed, deposition is most likely in areas of slack water, such as salt marshes, in bays where energy is dissipated in shallowing water, and where rivers meet the sea and clay particles **flocculate** (coagulate into bigger, heavier aggregates). Wind also has a role on the shore - transporting sand from beach to dune.

Features of deposition

Bayhead (pocket) beaches, such as Selwicks Bay at Flamborough or Lulworth Cove in Dorset, are **swash-aligned** and so curved to form outward facing arcs. This is due to wave refraction and the beach aligning itself towards their oncoming crests. Headlands protect these beaches on both sides. Consequently there is little input or output of sediment - the store is relatively static.

Open beaches, such as Bridlington and Filey, both on England's east coast, are **drift-aligned** and require a large supply of sediment in order to sustain them. Waves usually approach at an oblique angle, supplying and removing sediment in overall equilibrium. It is notable that both the examples above have northern protection from Flamborough Head and Filey Brigg respectively.

Bay barriers, such as Slapton Sands and Looe Bar, both on England's south-west coast, block the mouths of rivers which consequently have to drain through the sand or shingle - so forming beaches attached to both headlands.

The three examples above could be classified as **shoreline** as opposed to the **detached** beaches described next:

Curved spits require an abundant supply of sediment from longshore drift in order to build and sustain them. Indeed, they may demonstrate cyclic phases of construction and destruction. Blakeney Point in North Norfolk is a good example, but despite contemporary speculation as to its origins, Spurn Head, marking the southern extremity of the Holderness coast, is perhaps the most famous of all. Its history of repeated breaches and realignment, not least the building of six lighthouses, illustrates just how insecure such depositional features can be.

Double spits, such as at Poole in Dorset, most likely occur because longshore drift extends one spit in the direction of the prevalent winds whilst the other, encroaching on the estuary from the opposite direction, reflects the dominant winds.

NB: In all spits, periods of longitudinal growth tend to alternate with periods of lateral growth.

Tombolos are detached beaches such as spits linked to an island. For example, Chesil Beach links the Isle of Portland to the Dorset mainland. The location of the island must help in the formation of tombolos because waves will be refracted around the obstruction leading to deposition in the **wave shadow**. This, incidentally, is one of the arguments for why spits are so often curved at the end – the other being a change in prevailing wave direction.

Cuspate forelands, such as at Dungeness in Kent, are triangular outgrowths of shingle ridges formed by longshore drift from two opposite directions meeting to produce the sequence of ridges at right angles to each other.

Barrier islands are completely detached beaches 8-40 km offshore. The barrier forms offshore where waves break in shallow water on a gently sloping coast. Sea level has to be either constant or falling slightly in order that waves may build up the feature above water by constructive swash action. The coast of North Carolina, USA, and the North Sea coast of the Netherlands, are much quoted examples. However, other theories as to their origin have been suggested – that they are the remains of spits breached by storm waves or that they are sea bed deposits washed towards shore, again by rising sea levels.

Indeed, in many cases barrier islands have been observed to move toward the shore, but by **wash-over** of sediment from one side to the other. Some have eventually encroached on previously protected marshes and become shore features – a process which has been suggested as a possible origin to Chesil Beach.

Finally, sand dunes, mudflats, and salt marshes (saltings) must be considered:

Sand dunes are a product of aeolian, not marine processes. They require strong onshore winds and a wide foreshore in order to allow the sand to be dried out and blown in during low tide. A wide backshore with debris traps of shingle or vegetation is also required. Finally, salt-resistant and fast-growing plants such as marram grass are needed to anchor the dunes.

Mudflats and **salt marshes** form inland of spits and barriers and around estuaries. Mudflats are covered at high tide. Salt marshes, however, are only submerged by spring tides and are vegetated with, for example, spartina grass such as found west of Spurn Head. Salt marshes are drained by intricate networks of creeks which fill at high tide.

Changes in sea level

Eustatic and local changes in sea level are very important because sea level controls all coastal processes – waves, tides, currents, and so on.

Eustatic refers to world-wide variation normally associated with periods of glaciation. During an Ice Age, for example, water is 'locked up' as ice sheets and glaciers with consequent fall in sea levels. When they melt, sea level rises again. However, eustatic changes also reflect variations in the capacity of ocean basins. This is best illustrated in the Bay of Bengal where sea level rises by 1 m during the monsoon season due to excessive sedimentation from the River Ganges.

Local changes relate to tectonic and isostatic movements of the crust. For example, tectonic uplift of the Pozzuoli region near Naples has been significant. However, isostatic uplift of land in response to the removal glacial ice is more common.

Evidence of both lower and higher sea levels can be identified.

Lower sea levels: Before the Flandrian transgression, following the last Ice Age, Britain was joined to Europe – and Asia to Alaska. The remains of mammoths and sabre-toothed tigers are found in Britain because they could not move south once the English Channel flooded. There are submerged forests in the Dogger Bank fishing grounds of the North Sea, submarine river systems in the Indian Ocean, and drowned spits, dunes, and drumlins elsewhere.

Higher sea levels: Ancient wave cut platforms as current cliff tops, raised beaches, caves, degraded cliffs with wave cut notches, fossil stacks, shingle ridges – all at or around King's Cave on Arran suggest, in this case, isostatic uplift of the land being faster than rising sea level associated with the Flandrian transgression.

CLASSIFICATION OF COASTS

Various coastal classifications have been published based upon various criteria such as:

Changes in sea level (D.W. Johnson, 1919)

1. **Submerged coasts** are the result of a positive change in base level, either because the sea has risen, or the land has fallen – both resulting in a transgression. Such submergence causes great indentation with, for example, fjords, rias (drowned river valleys), and islands (the tops of drowned hills and headlands) associated with submerged upland areas, and less obvious, but nonetheless important, drowned flood plains, such as the Humber Estuary, associated with lowland areas.

2. **Emerged coasts** are the result of a negative change in base level, either because the sea has fallen or the land has risen – both resulting in a regression. Such emergence leaves far straighter coastlines, with raised coastal features such as those found on Arran (discussed previously).

3. **Stable** or **neutral coasts** either show no sign of changes to the level of the land relative to the sea or are those produced by non-marine processes such as river delta or volcanic island formation.

4. **Compound coasts** exhibit any mixture of the three above. For example, Sussex has a flat coastal plain and smooth shoreline with raised beaches – yet is currently being submerged.

Criticisms of the arguably simplistic nature of this classification have led to other systems. Yet any classification, by definition, generalises by organising knowledge more efficiently in order that understanding may be promoted. For example R. Valentin (1952) suggested a basic **advancing-retreating** distinction; F.P. Shepard (1963) **primary** or **secondary**; A.L. Bloom (1970) **hard rock** or **soft rock**; J.L. Davis (1980) **high energy**, **low energy** or **protected**. Suffice it to say that no single classification is beyond criticism and so there is much to be said for using whichever system best clarifies one's understanding.

8

SOILS

Soils are of special significance because they represent the link between a lifeless world of rock and rock waste and the living world of plants, animals, and humanity. Clearly they integrate organic and inorganic elements of the environment. Soils may be viewed as the earth's most important natural resource, effectively irreplaceable if lost to erosion.

Pedology (soil science) studies the pedosphere which one could argue as central to an environmental system connecting atmosphere, biosphere, hydrosphere, and lithosphere.

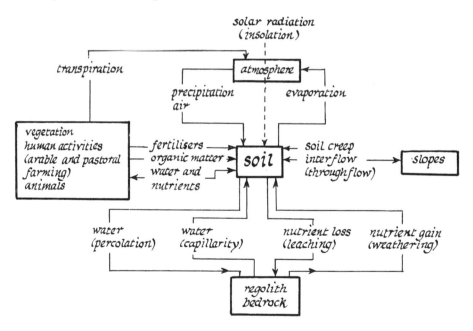

There are various definitions of soil from the trite – *'anything plants will grow in'* – to the more formal – *'a mixture of particles of weathered rock, decayed organic matter, water, and gases in which living organisms are active.'*

Much of the material in soil is derived from rocks which are continually being broken down into small stones and tiny particles of mineral matter known as **regolith**. (Strictly speaking this is a collective name for all loose, weathered material above the bedrock – therefore including soil. However, it is usual to regard soil as distinctive – with regolith as its parent material.)

Decayed organic matter derived from plants and animals is added to this rock waste in the form of rotted leaves, roots, and excreta. This is called **humus**. The amalgam that is soil also contains gases (air) and water in addition to living plant and animal organisms, including bacteria, fungi, and earthworms, known as **biota**. It is notable that a soil should be capable of holding moisture within itself against the force of gravity. Moisture is both evaporated directly from the soil into the atmosphere, and indirectly via transpiration from vegetation. Indeed the relationship between precipitation and evapotranspiration determines to some extent the type of soil that eventually develops. (Soil water and various nutrients in the soil support the growth of vegetation and this of course supplies food for animals and ourselves.) Water may infiltrate through the soil into the subsoil and then percolate down to the parent rock - especially if precipitation is greater than evapotranspiration. This process can cause nutrient loss by **leaching**. On the other hand, parent rock weathering deepens the soil and provides extra nutrients. Nutrients may also be added to the soil, in certain climatic conditions, by soil water rising through capillary action.

INFLUENCES ON SOIL DEVELOPMENT

1. **Parent material** is most important in young immature soils. Its influence decreases, however, as the soil becomes older.

2. **Relief** is important because angle of slope affects runoff and vulnerability to soil erosion. Likewise other site characteristics such as **altitude** can affect climate, just as **aspect** influences solar warming.

3. **Climate** is of major importance. Precipitation, temperature, and their seasonal and diurnal variations all affect soil formation.

4. **Biota**, although usually dependent on climate, can act as an independent variable. This is because the supply of organic material can be altered and interrupted should the vegetation be changed. Also organisms such as bacteria, fungi, and earthworms have a marked effect on soil formation by helping the breakdown and incorporation of organic material.

5. **Time** is always relevant because different soils form at different rates. Thousands, sometimes millions of years may be required to develop a soil. Needless to say maturity comes with age.

6. **Human activities** clearly influence the characteristics of the soil. Ploughing and draining, for example, affects the soil structure and horizon arrangement. Natural plant cover is removed and replaced with built-up areas, forests, fields, and pasture. Consequently, water and nutrients in the soil support crops and domesticated animals. Additionally, farming may involve the return of organic matter to the soil and renewal of soil nutrients by the application of artificial fertilisers.

Soil composition

Soil contains solid materials of both inorganic and organic origin. It also contains various liquids and gases.

Inorganic material may consist of particles of various sizes from pebbles, through sand and silt to fine-grained clay. Directly weathered sand grains (silica), for example, will show no chemical change. However, clay may result from chemical weathering of aluminium silicate minerals. Parent rock weathering will also supply the soil with compounds of sodium, potassium, calcium, and magnesium. These **cations (bases)** are important plant nutrients.

Organic material consists primarily of dark brown or black humus. This is derived from plants and animals that have been partly decomposed by detritivores and saprophytes (see *Biogeography*) such as earthworms, bacteria, and fungi. Plant leaves and stems on the surface are mixed into the soil by various organisms - mainly earthworms, but moles, mice, rats, and rabbits too. Decaying plant roots on the other hand are already incorporated. The amount of humus will depend primarily on the amount of vegetation that the soil supports. However, it also depends to some extent on the climate. High temperatures, for example, favour rapid humus production. Humus quality will depend on vegetation type with some plants and trees requiring a large quantity of nutrients to build stems, leaves, and root systems. These are eventually returned to the soil as **mull** humus which is rich in plant nutrients. This kind of **nutrient cycle** is very common, indeed characteristic of grassland and deciduous woodland. Other types of vegetation, such as coniferous evergreen trees, take up fewer nutrients and in turn return less. This poorer humus is known as **mor**. The presence of humus gives soil a brown or black colour. (Iron gives a red or yellow colour. However, in poorly drained soils, iron compounds suffer reduction due to lack of oxygen resulting in green, grey, or even blue coloration.)

Soil water (soil solution) exists in the interstices of the soil and can move through it. Supply is by precipitation – exit by evapotranspiration, interflow (throughflow) to river channels, and percolation down to the water table.

Gases, entering mainly from the atmosphere, occupy the interstices not occupied by water – therefore they cannot enter a fully saturated soil. **Soil air** is 'richer' than the atmosphere – containing less oxygen, but more carbon dioxide and methane, for example. Of great relevance to the decomposition process, by which plant and animal matter is transformed into humus, is that many types of bacteria can only exist where air is present.

Soil texture

Texture refers to the size of the individual particle constituents of a soil. Three broad distinctions are made – sand, silt, and clay. A **loam** is a soil whereby a mixture of these particle sizes are found. We refer, for example, to a 'clay loam' or 'silt loam' and so on, depending on which particle is dominant.

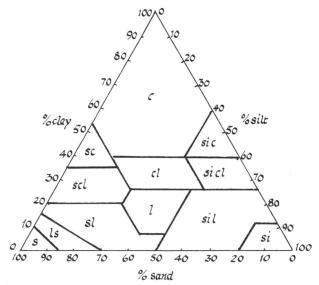

c clay   l loam   s sand   si silt   cl clay loam   ls loamy sand
sc sandy clay   scl sandy clay loam   sl sandy loam   sic silty clay
sicl silty clay loam   sil silt loam

Texture is related to parent material. Sandstones, for example, produce sandy material - shales, silty material. Acid igneous rocks rich in quartz, like granite, tend to produce sandy material, whereas basic igneous rocks, like basalt, tend to produce clay. Soils may also contain tiny particles called **colloids** that can be composed of either mineral or organic matter. Soil texture determines to a great extent its permeability. Water rapidly infiltrates though a sandy soil, whilst a clay soil retains it in its tiny interstices. Clay soils feel sticky when wet, silty soils feel smooth, and sandy soils gritty. Soil permeability is also influenced by pore spaces created by worms, burrowing animals, and the decay of plant roots.

Soil structure

Structure refers to the way grains of soil adhere together to form larger pieces called **peds** or lumps. The soil grains are held together by colloids.

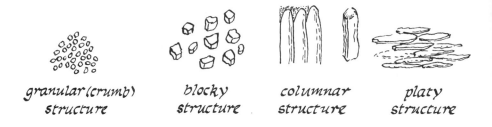

granular (crumb)   blocky   columnar   platy
structure   structure   structure   structure

Various structures can be identified in relation to the size and shape of peds. In a **granular** or **crumb** structure grains adhere together to form rounded pieces. A **blocky** structure occurs when larger aggregations occur with sharp corners and irregular shape. **Columnar** structure is characterised by vertically elongated prismatic columns. **Platy** structures occur when the soil grains form flattish horizontal aggregations.

The clay-humus complex

Soil is not simply a mixture of sand, silt, and clay with humus. Humus tends to be linked with clay particles to form the **clay-humus complex**. Here tiny particles (**micelles**) of clay and humus can hold plant nutrients (cations or bases) in such a way that they are available to plant roots. If the clay-humus complex were not present, rain water would be able to wash soluble plant nutrients downwards through the soil.

Soil acidity

Some soils are naturally alkaline   such as recently developed soils on chalk or limestone bedrock. However, many soils tend to be acid because the cations of the clay-humus complex tend to be replaced by H_2 from the soil solution.

<7   7   >7
acid   neutral   alkaline

A slightly acid soil (pH 6.5) is usually best for plant growth since many nutrients are then most soluble and so easily available to plants.

Soil processes: cation exchange and the soil profile

Cation exchange takes place in two forms:

(i) Direct contact between plant and clay-humus complex.
(ii) Contact between soil water and clay-humus complex.

The clay-humus complex is fundamental to the various processes that take place in soils. It holds the largest reserve of plant nutrients in the form of cations (bases) of sodium, potassium, calcium, and magnesium. Grains of purely mineral sand and silt can supply no nutrients to plants. Soil colloids are electrically charged and can attract cations from the soil water (solution). These nutrients become attached to the micelles of the clay-humus complex. If this clay-humus complex did not exist, these nutrients would be leached out of the soil by percolating water. When the plant roots come into contact with the clay-humus micelles they give off hydrogen ions in exchange for plant nutrients. So, in effect, the cations of the clay-humus complex are replaced by H_2 ions. The cations can be replaced, subsequently, by creation of further humus by the recycling of plant debris.

Cation exchange also takes place between the clay-humus complex and the soil solution. H_2 ions from the soil solution can replace the cations on the surface of the soil colloids, thus causing an increase in the level of the soil acidity. When the clay-humus complex is almost completely saturated with H_2 the soil is acidic (pH 4). Whilst the clay-humus complex is stable when saturated with cations, as these are replaced by H_2 ions it tends to become increasingly acid and unstable. Eventually the clay-humus complex may disintegrate and be leached away. The most extreme form of this will leave a purely mineral soil. The application of fertilisers, however, can have the effect of replacing the cations from the clay-humus complex and so prevent a decline in soil fertility.

The soil profile (solum) refers to the variations that occur in the characteristics of a soil as you go down from the surface towards the underlying bedrock. A well developed soil tends to have horizontal layers (**horizons**) which differ in texture, colour, and chemical composition. These horizons, created by processes operating within the soil, are transitional between mineral material at depth and organic matter at the surface. The surface layers comprise of organic matter, derived mostly from the vegetation. This accumulates in the A_o or O horizon and is usually divided into three layers:

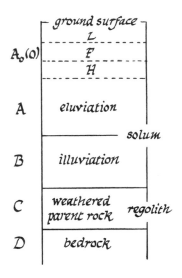

L Litter: loose leaves and
                    raw organic debris.
F Fermentation: active decay and
                    partial decomposition.
H Humus: decay more complete –
                    dark humus combines
                    with clay fraction of
                    the soil to become the
                    clay-humus complex.

In the A and B horizons, the mineral content of soil increases downwards towards the bedrock.

In a humid temperate climate, the soil solution generally moves downwards through the soil, since precipitation is greater than evapotranspiration. Therefore in the A horizon (**zone of eluviation** or outwashing) leaching occurs and soluble soil constituents are carried downwards. H_2 ions from the soil solution replace cations in the clay-humus complex and the nutrients are carried away - thus increasing soil acidity. The clay-humus complex may even disintegrate and be entirely leached out of this **topsoil**.

The underlying B horizon (**zone of illuviation** or inwashing) generally has a smaller humus content and a greater proportion of weathered rock material. However, nutrients leached from the A horizon may accumulate in this **subsoil**. The A and B horizons represent **true soil**, whilst the C horizon is weathered parent rock and D unaltered bedrock.

Soil classification

Classification is necessary if we are to organise our knowledge in a systematic way. By classifying we can come to understand broad generalisations in addition to detailed individual descriptions or explanations. Classifications do not add to knowledge - their value lies in the fact that they organise knowledge more efficiently - despite the appearance of anomalies at times.

The classification of soil types most used is the **zonal system**. This subdivides the world's soils into three major categories - zonal, intrazonal, and azonal.

1. **Zonal soils** have characteristics that tend to reflect climatic conditions. They are created over a long period of time by this influence and of the associated vegetation. They are well drained and develop on gently undulating surfaces. Due to their maturity, parent rock has limited influence.

However, climate as the main factor leads to criticism in that it can be argued that considerable climatic changes may have taken place during the period in which zonal soils have evolved. Also there can be many small-scale climatic variations within quite a small area. Nevertheless, this zonal soil concept allows us to describe and explain in relatively simple terms the broad features of the latitudinal distribution of soil types over the world and to relate them to the climatic climax vegetation. It is not possible, however, to study local, small-scale soil variations in terms of zonal soils.

TUNDRA SOIL

These soils form in tundra areas of permanently frozen subsoil (permafrost). Cool summers dictate a short growing season. The soil is frozen in winter and waterlogged in summer, because the water cannot percolate downwards through the permafrost. Most of the mineral material is frost-shattered rock fragments contained in a blue/grey mud. It is in this matrix that **gleying** occurs whereby ferrous oxides are reduced to blue or grey ferrous salts in the absence of oxygen due to waterlogging. The surface layer is often peat because bacterial activity is so weak that humus development is very slow. There are no clear soil horizons because alternate freeze-thaw cycles mix the underlying mud up through the layer of peat.

PODSOL

This is the zonal soil of the taiga (boreal coniferous forest of the northern hemisphere). It is also found in other areas with cool climates such as in the UK on heathland and moors, and in areas of sandy soil such as fluvioglacial outwash plains. Podsols are often associated with coniferous evergreen trees such as pines. Areas of boreal forest do not receive particularly heavy rainfall, but the podsolisation process requires a general downward movement of water through the soil. Precipitation, therefore, exceeds evapotranspiration because of the low temperatures. Also, forest vegetation influences the moisture balance - coniferous trees cast a heavy shade over the ground and shelter it from drying winds. Consequently, moderate precipitation can easily provide a surplus over evapotranspiration and so allow downward percolation.

The podsol has a very poor nutrient cycle. Coniferous evergreens do not take up cations such as calcium, magnesium, and potassium and so these nutrients are not returned to the soil when the leaves fall - hence a poor mor (acid) humus.

The A_o horizon comprises of relatively unaltered leaves (needles) which are increasingly decomposed downwards through the F and H horizons.

The A horizon immediately below the H horizon is dark grey due to staining.

Most, however, is of light grey, almost white, sandy material. All colouring material - humus and iron for example, has been leached out of this zone of eluviation.

	coniferous trees
A_o	L thin F protracted H mor
A	light grey sandy
	indurated layer (iron pan)
B	red/yellow/ brown staining
C	transition to bedrock

The B horizon is darker and denser in texture due to accumulation of clay, humus, and aluminium leached from above. Indeed, within this zone of illuviation, iron may accumulate to form an orange-coloured **hard (iron) pan** marking the highest point of the water table. The B horizon merges into the parent rock.

The clear differentiation of horizons in a podsol indicates that there may be few mixing agents such as earthworms.

BROWN EARTH SOIL

These soils are associated with deciduous forests of wetter and warmer climates which allow a longer growing season than coniferous trees. Deciduous trees also demand and supply more nutrients. Deeper roots draw cations from deep within the soil - even from the regolith. A regular supply of nutrients is returned, however, to the soil surface by the annual shed of leaves.

Soil fauna is much more abundant - intensive bacterial activity, earthworms, and rodents break down and mix plant debris very efficiently.

Generally leaching is not so severe unless precipitation is particularly heavy or the parent rock has few cations. However, some leaching of the upper part of the profile is natural with clay and humus leached downwards - but horizons are not clearly differentiated because of the mixing just described.

The A_o horizon comprises a thick litter of leaves with dark brown, slightly acid, and very rich mull humus beneath.

The A and B horizons are difficult to differentiate but pale as you go down because humus becomes less abundant.

		deciduous woods
A_o		*L thick* *F efficient* *H mull*
A		*crumb structure* *well aerated* *some staining* *from humus*
B		*no horizon* *differentiation*
C		*transition to* *bedrock*

CHERNOZEM AND RELATED SOILS

These soils are associated with temperate grassland vegetation in warmer and drier areas. (However, in continental interiors winters may be very cold with much frost.) Soil moisture evaporates freely and there is considerable transpiration loss from grasses. Therefore, precipitation does not easily penetrate deeply and sometimes soil water can move upwards. The soil usually freezes in winter, preventing leaching.

The nutrient cycle is very rich - grassland takes up more cations than deciduous forest, hence more humus is produced to give a richer soil. Consequently, the soil is dark in colour. Indeed, the deep, fertile, moisture-retaining crumb structure of this 'black earth' soil makes it ideal for cultivation. The A horizon is very rich because biota mixes humus into the soil quickly. Its black/dark brown colour gradually lightens to a yellow/brown colour beneath - with a recognisable B horizon unlikely. However, because chernozems contain a considerable amount of calcium carbonate, this tends to be deposited in the burrows of the soil biota, forming a layer of nodules at the maximum depth to which rainfall penetrates. (The presence of calcium carbonate in the soil counteracts any tendency for clay compounds to break up and for leaching to begin.)

Related soils include the prairie, chestnut brown, and serozem:

Prairie soils form under the same (grassland) vegetation but in areas of greater precipitation. There can be no layer of calcium carbonate concretions therefore.

Chestnut brown soils are associated with poorer grassland due to drier conditions. There is consequently less humus - hence the paler colour. Also there will be less leaching and the layer of calcareous concretions will be shallower.

Serozems are associated with even sparser vegetation found at desert margins. The low humus content gives the soil a greyish coloration. Its high calcareous content is brought towards the surface by upward movement of soil water provoked by the hot, dry conditions. This soil is potentially rich in mineral nutrients, hence may be very fertile if irrigated.

TROPICAL RED SOILS (LATOSOLS)

These soils are associated with the tropical rainforest. The rapid nutrient cycle reflects demanding forest growth yet constant return of nutrients via, for example, leaf fall. The hot, moist climate provokes rapid decomposition of all organic material. Consequently, the soil is not very acid and, paradoxically, infertile because the nutrients are not long lasting - hence the inadvisability of forest clearance for agricultural use.

The soils are well drained. Consequently silica and clay colloids may be removed by leaching, leaving sesquioxides of iron and aluminium (Fe_2O_3 and Al_2O_3) in the surface soil. Should these become concentrated a lateritic surface crust may develop, but this is unlikely if the rainforest cover is maintained.

The soil's red colour comes from compounds of iron.

Zonal soils may be further subdivided into two groups:

Pedocals are soils, such as chernozems and chestnut browns, in which calcium carbonate tends to accumulate in generally drier climates with less leaching.

Pedalfers are soils, such as podsols and tropical red 'latosols', in which iron and aluminium oxides tend to accumulate in wetter climates where leaching is more intense. They are characterised by a layer composed of iron and aluminium sesquioxides.

2. **Intrazonal soils** reflect the dominance of other factors, such as the characteristics of the parent rock rather than climate and related vegetation. This is because within any climatic zone there can be very great variations in relief, rock type, and drainage conditions – all of which can have a great influence on soil type. A limestone bedrock such as at Malham, for example, gives rise to alkaline (terra rossa) soils in an area where the climate might be expected to produce leached podsols. It is possible, therefore, for a particular type of intrazonal soil to occur in different regions which have different climates simply because of the dominating influence of, say, parent rock.

Three main types of intrazonal soil are distinguished according to the predominant influence upon their characteristics. These are hydromorphic, calcimorphic, and halomorphic.

HYDROMORPHIC SOILS

These soils develop in areas of excessive water – usually in swampy areas (**bog soils**) and low-lying river flood plains (**meadow soils**).

Their distinguishing characteristic is that part of the profile is constantly saturated with water. The level of the water table can vary from season to season. Consequently, even the upper parts of the soil are periodically saturated.

Often, large amounts of organic matter, in the form of plant remains, accumulate because too much water tends to restrict bacterial activity which could transform it into humus. Therefore, the vegetation is only partially decomposed, leaving an upper horizon of peat.

(**Bog peat** forms in poorly drained upland areas and is extremely acid. In lowland areas of fen, however, the presence of river water containing calcareous material can lower its acidity.)

Saturation is clearly the key to understanding hydromorphic soils. The lower horizons are likely to be saturated for longer periods. Oxygen, therefore, will be rapidly used up by micro-organisms, resulting in gleying which will become more intense at greater depth. Iron compounds will reduce chemically by giving up their oxygen, from a ferric to ferrous state - hence changing from a red to green/blue colour. Other minerals will produce a grey colour. Only where oxygen is made available from air or percolating water, such as along root channels or in small fissures, will oxidation occur - causing red or brown mottles and streaks to interrupt the overall bluey grey coloration so indicative of gleying.

However, it is also possible for gleying to take place much nearer the surface. Where the soil consists of relatively impermeable boulder clay, for example, there may be no identifiable water table. Consequently, it may become saturated from time to time causing gleying at a shallow depth and hence a **surface water gley soil**. Also gleying is not restricted to hydromorphic soils. Podsols may be mottled through gleying beneath the hard pan.

CALCIMORPHIC SOILS

These come in two versions - rendzinas on soft chalk and terra rossas on hard limestone.

Rendzinas are shallow soils associated with humid climates on softer chalks. They are rich in humus, hence dark in colour, and only slightly alkaline. Normally the illuvial B horizon is absent, giving rise to the name 'A-C soil'. This is due to the absence of leaching in the A horizon. Also, the underlying weathered chalk is mixed upwards into the soil.

Terra rossas are associated with harder limestone which inhibits weathering and so reduces upward mixing. Consequently the soil has a lower pH, allowing clay minerals to be broken up to produce the ferric oxide which gives it its distinctive red colour.

HALOMORPHIC SOILS

These are associated with arid climates and are created by the concentration of various soluble salts in particular localities. There are three broad types:

Solonchaks have the greatest concentration of salts such as sodium chloride, gypsum, and calcium carbonate. These may rise by capillarity if the water table is high enough and so create a surface crust. Alternatively, the salts may be carried in solution by surface runoff, which collects in low-lying areas, only to be precipitated as the water evaporates. A solonchak can also form as a result of irrigation with insufficient drainage - salts are precipitated from the irrigation water.

Solonetz are thin grey soils overlying a dark-coloured alkaline layer. They have a smaller concentration of salts than solonchaks for two possible reasons - a lower water table, hence less capillarity, or greater leaching because of heavier rainfall.

Soloths are degraded forms of the solonetz - similar but more fully leached, with a lower salt content. This may be due to higher rainfall or perhaps a greater supply of irrigation water combined with adequate drainage.

3. **Azonal soils** are generally immature and skeletal, with poorly developed profiles. These young soils are made up almost entirely of fragments of mineral material with minimal humus. They are affected little by the soil-forming processes associated with zonal and intrazonal soils and do not have well-developed profiles because of their recent development. There are three groups - lithosols, fluvisols, and regosols.

Lithosols comprise coarse-calibre material such as sand or gravel from screes and newly created glacial moraines. They often occur on slopes where surface runoff of precipitation is common because this washes the finer material downslope. Little vegetation can grow in such conditions, hence there is minimal humus. Lithosols are very likely to be affected by mass movement.

Fluvisols comprise relatively recently deposited alluvium in or near river flood plains. Often grass vegetation supplies some humus. Where there is a fairly high water table, gleying is common - hence fluvisols are common on flood plains. River terrace soils, however, are usually better drained and will probably have existed for a longer time - hence they are more likely to have developed some of the characteristics of zonal soils.

Regosols are soils consisting of fairly deep layers of relatively fine rock material such as on sand dunes, wind-deposited loess, and glacial and volcanic deposits.

Soil catena

This is a sequence of various types of soil profile that occur in succession from the crest to the foot of a hill slope. The term **catena** is usually restricted to hill slopes with a uniform type of underlying rock and experiencing no marked climatic variation. Therefore, the sequence of profiles must be related to the characteristics of the slope itself. Variations in soil type along the slope are mainly the result of variations in soil drainage which is faster on steep slopes and slower on gentle slopes.

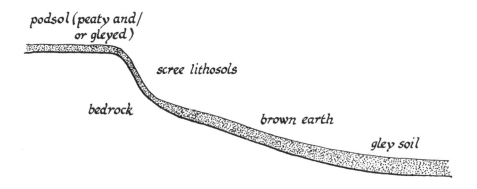

An upland area in Britain, for example, might illustrate a zonal soil such as a **podsol**, directly related to the cool, moist climate on a summit plateau. Low temperatures limiting bacterial activity may restrict humus development to peat, and drainage may be poor on this level surface, leading to a gleyed horizon low in the profile. **Scree lithosols** may occur beneath free faces on the slopes edging the plateau. Certainly steeper slopes lead to thinner soils (and *vice versa*) and, with better drainage, soil nutrients are likely to be leached downslope to accumulate in the soil near the base of the slope, thus forming a less acid **brown earth**. Finally, because the soil's moisture content will increase as you go down, the level ground at the foot of the slope may allow water to accumulate, causing gleying, and so perhaps forming an intrazonal hydromorphic **gley soil**.

Soil erosion

Soil erosion is normally blamed on our removal of natural vegetation cover, so leaving the ground directly exposed to the elements. However, replacement of this vegetation by planned land-use, such as appropriate agriculture, can and does preserve soils – albeit in altered states. Where poor land management, often provoked by population pressure, has allowed overgrazing, overcultivation, reduced fallow periods, and, especially, deforestation – erosion is inevitable. Indeed, in many **economically less developed countries** (ELDCs) erosion relating to these factors is now reaching alarming proportions. However, **economically more developed countries** (EMDCs) are not immune to the forces of wind and water removing fertile topsoil, so reducing soil thickness and hence room for roots, especially where deep ploughing up and down slopes and monoculture is practised on large fields.

The catastrophic **wind erosion** causing the infamous dust storms of the early 1930s in the American Mid-West was not, as often misinterpreted, the sole result of sequential droughts, but due to removing the natural grassland vegetation and leaving the soil exposed for long periods between crops. Wind erosion can often be seen today in south-eastern arable farming counties of England, due partly to the removal of hedgerows to increase cultivable area and mechanical efficiency.

Water erosion on slopes is a particular hazard and takes different forms:

Sheet erosion (when water moves slowly, but evenly over the surface) may be so insidious that it may not be recognised until too late, although the thicker topsoil down slope may not be unwelcome. Faster flow, however, is less likely to be missed because the water begins to cut into the surface causing **rill erosion**. Even larger volumes and higher speeds cause **gully erosion** which may result in devastation of unconsolidated soils such as wind-deposited loess. Material removed by rill and gully erosion' may choke river channels and 'silt up' reservoirs and harbours, necessitating dredging. **Rainsplash** is also an important erosive agent – not just because it causes material to be splashed upwards and outwards, but by sealing and 'puddling' the surface, so reducing infiltration and promoting erosive runoff. Thus the protective role of vegetation can never be underestimated whenever conservation measures are proposed.

Soil conservation

Soil conservation, therefore, in its simplest form is vegetation cover – and is essential on steep slopes by means of forest, orchard, or pasture. But this is not always possible or desirable on gentler slopes where arable cultivation is practiced. **Terracing** hillsides represents an ancient, yet nonetheless invaluable method of conserving soil in grape, olive, and particularly rice cultivating areas. **Contour ploughing** allows no furrows to be exploited by running water and, like terracing, can promote infiltration and moisture retention. However, it does increase the risk of accidents caused by machinery tipping over. **Crop rotations** reduce the likelihood of soil exhaustion, and failed harvest, by maintaining soil quality. They need not involve expensive chemical fertilisers because leguminous species such as clover, beans, and peas have long been recognised as 'soil improvers' by fixing, and so adding, atmospheric nitrogen. **Strip cropping** is more sophisticated – whereby two or more crops are grown along the contour, preferably using biennials or perennials in the alternate strips so that any erosion of a bare strip is checked by the next. **Shelter-belts (wind-breaks)** of trees reduce wind erosion further. **Afforestation**, however, represents the best long-term solution to soil degradation through interception of rainfall, shading from the sun, and root binding. **Controlled grazing** reduces the number of animals – and so their need, through inadequate pasture, to eat down to the bitter root, killing the grass and exposing the resulting bare soil to trampling. However, this may not always be easy to implement, especially in regions such as the Sahel of Africa where overpopulation, and persisting cultural traditions such as bride dowries, emphasise herd quantity rather than quality. Indeed, a perverse benefit of repeated prolonged droughts in both the 1970s and 1980s has been cattle and goat depopulation, allowing restocking with smaller but better quality herds. This region also illustrates a final, yet pertinent point regarding soil conservation measures. They need not be expensive or elaborate. The much quoted Burkina Faso **stone lines** lain along contours (determined by matching water levels in the upturned ends of a length of filled hosepipe) have reduced runoff on gentle slopes, so significantly increasing infiltration and, consequently, crop yields.

9

BIOGEOGRAPHY

Biogeography is concerned with understanding and explaining the distribution of the multitudinous forms of plant and animal life which inhabit the biosphere. The **biosphere** is the biologically inhabited parts of the lithosphere, hydrosphere, and atmosphere and is often thought of as *'the living envelope of the planet.'* An appreciation of ecological systems (ecosystems) is central to biogeographical studies.

Ecosystems represent *'an ordered and highly integrated community of plants and animals together with the environment that influences it.'* They must function as complete units and be self-sustaining, with all organisms co-existing in equilibrium.

Every ecosystem has four components in equilibrium, representing the so-called *'balance of nature.'* Whilst inorganic substances are **abiotic** (non-living), producer, consumer, and reducer organisms are **biotic** (living).

Inorganic substances include the minerals in the soil matrix which were released by weathering of the bedrock. Other elements, such as nitrogen and carbon, are dissolved in rain water and enter the soil, as do the gases which make up the air (mainly oxygen). All of these provide inorganic nutrients for the producers.

Producer organisms (self-feeding **autotrophs**) are the green plants containing chlorophyll. **Photosynthesis** is the key process in their growth whereby energy from sunlight (**insolation**) and the chlorophyll convert carbon dioxide and water into their organic compounds - the so-called 'building blocks' of plants such as carbohydrate and tissue. Oxygen, essential to supporting life on the planet, is released as waste. They are called **primary producers** because they are the first organisms in the food chain to create organic food and so store nutrients. They also use inorganic nutrients from the soil which are absorbed by cation exchange and **active transport** - drawn in solution through the roots.

Consumer organisms are all animals including birds, fish, reptiles, and insects. We distinguish between primary consumers, which are vegetarian (**herbivores**), and secondary consumers, which are meat eaters (**carnivores**).

FOOD CHAINS AND ECOLOGICAL PYRAMIDS

Food chains, with each link known as a **trophic level**, express the progressive transfer of energy through the ecosystem. For example:

Producer	**Primary consumer**	**Secondary consumer**	**Tertiary consumer**
grass	rabbit	fox	flea

Many of the nutrients consumed are progressively lost because of the various animal functions. Energy is lost in respiration and movement, whilst nutrients are released in defaecation. Consequently, energy from the sun is progressively lost along the chain.

(More complex progressions are called **food webs**.) Therefore, we could feed far more people on, say, cereal, than on meat from the cattle or pigs that ate it. The **biomass** (total weight of living organisms – both plants and animals) is of great relevance here. **Ecological pyramids** show the relative proportions of biomass required to support the next trophic level. They have been used to great effect in demonstrating the extravagant food and energy wastage, in terms of processing, involved in diets found in EMDCs compared with ELDCs.

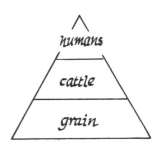

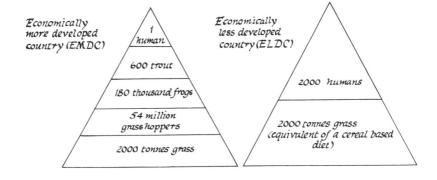

Reducer organisms (decomposers) fall into two groups:

(i) **Detritivores** eat dead and decaying organisms. Examples include lice, earthworms, and vultures.

(ii) **Saprophytes** cause the decay and breakdown of dead organisms and excrement into mineral salts and carbon dioxide which are recycled, as nutrients, back into the system for primary producers to use. Examples include bacteria and fungi.

In any study of an ecosystem, nutrients should be thought of as being recycled whilst energy is lost. Examining structure and function represents a particularly ordered approach:

Structure is the shape or morphology of the system. This is best shown by diagram in order that the four components described earlier are clearly illustrated.

Functioning explains the links between these components - the amounts of energy and nutrients, and the interrelationships between them. It would include seasonal or long-term changes to the system.

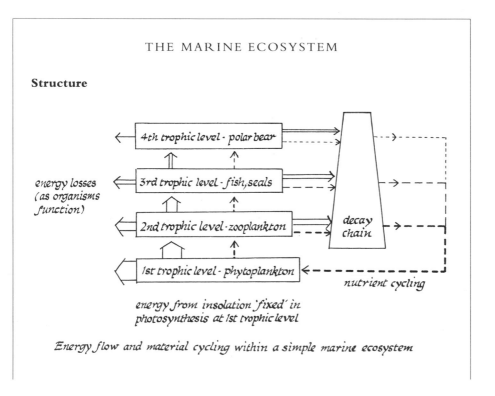

Energy flow and material cycling within a simple marine ecosystem

Functioning

The sun is the energy source upon which all life depends. It enters the biosphere through photosynthesis, in this case of phytoplankton. These are the primary producers which form the first trophic level. The second trophic level consists of zooplankton (herbivores) which eat the phytoplankton. The third trophic level consists of fish and seals. These are carnivores which eat herbivores. Likewise, the fourth trophic level also consists of carnivores such as polar bears. Thus energy is passed on down this simple linear food chain, or into the food web should it be more complicated. Some energy is lost along the chain because organisms die and decompose, so entering the decay chain instead. The decomposers break down the material into its constituent gases, minerals, and water before returning them to the system for recycling. This process is called **nutrient cycling**. Further energy is lost in the functioning of organisms. Indeed, only a small percentage is 'locked in' to be passed on.

When the basic requirements for life (water, oxygen, nutrients, heat, light, and circulation - for nutrient cycling) are considered, it can be seen that the sea provides more favourable conditions for organic production than the land. In the sea there is no water shortage, with abundant oxygen and carbon dioxide readily available. The sea is a **nutritious solution** - its salinity having been caused by dissolved minerals - again, readily available. Temperature variations are less marked than on land. Consequently, organisms do not have to cope with such a range. The transparency of the sea allows a thicker photosynthetic zone than on land. Finally, circulation is variable according to latitude. It is best in the subarctic waters of temperate latitudes where marine organic production is highest. Here, arctic and tropical waters meet with much mixing. Seasonal variations increase the circulation which is greatest over continental shelves, subject to wave action churning up the bottom water. Also, where warm surface currents diverge (separate) as, for example, along the equator, the bottom water upwells. Upwelling also occurs where offshore currents and winds carry surface water away. For example, offshore currents off Peru and westerly winds off Newfoundland both allow nutrient-rich bottom water to upwell, so producing further organic matter. Finally, the higher frequency of gales and consequent water turbulence in subarctic and temperate latitudes further increases this trend of increased circulation. Tropical waters and the Doldrums, in contrast, display poor circulation because warm waters remain at the surface. Although there is plenty of light, and high temperatures for unchecked growth, there is very little mixing of deep water and so poor nutrient recycling. Polar waters, likewise, are very stable, with, therefore, limited nutrient cycling and, given even lower temperatures, less organic production. This is because solar radiation is reflected off the ice, and cold water is denser, therefore sinking to the bottom, taking debris and nutrients with it.

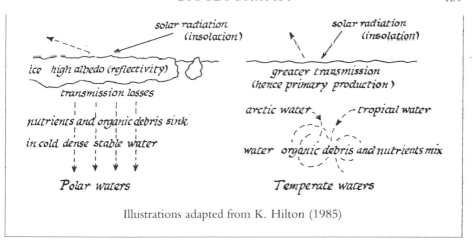

Illustrations adapted from K. Hilton (1985)

The marine ecosystem is a relatively simple one. Ecosystems on land, however, tend to be more complex. This is because there are more variables involved. Vegetation composition depends upon the interaction of the various elements of the environment in which it lives - its **habitat**. Water, temperature, wind, light, relief, soil, and biotic factors all influence ecosystems on land.

1. **Water** availability is critical. Therefore, the nature, amount, and seasonal distribution of precipitation is important. Water absorption by the roots, its distribution within the plant itself, not least the ease of passage through leaf pores (**stomata**) are further influenced by temperature and humidity conditions. Temperatures, for example, may be too low for **osmosis** - the way in which roots take up water. Consequently, plants evolve with characteristics appropriate to their specific habitats:

- **hydrophytes** - such as water hyacinth, in water,
- **hygrophytes** - such as sedge, in wet, marshy conditions,
- **halophytes** - such as spartina, in saline conditions,
- **mesophytes** - such as oak, where there is neither water excess nor deficiency,
- **xerophytes** - such as cactus, in arid conditions.

2. **Temperature** range is important in that any plant will be tolerant to set maximums and minimums - the latter usually around 6°C. The optimum for growth will be within these extremes. Temperature also affects evapotranspiration and, as indicated above, osmosis rates.

3. **Wind**, likewise, affects evapotranspiration rates, local temperatures, and humidity. It disperses seeds and, if strong enough, damages plants.

4. **Light** is needed for photosynthesis. The amount and intensity will depend upon local aspect and relief, proximity of other plants, cloud cover, season, and latitude.

5. **Relief** affects micro-climates by influencing shelter, and if high, inducing precipitation. Certainly, slopes have great bearing on the drainage, texture, and depth of soil.

6. **Soil** conditions are very important – not just the characteristics above, but the organic and mineral content. Deficiencies of nutritious elements such as calcium, nitrogen, and potassium may restrict growth just as excesses of, for example, aluminium are harmful.

7. **Biotic factors** include soil organisms mixing and aerating the soil, animals grazing, and birds spreading seeds. Human activities, typically, are particularly influential given widespread clearance of natural vegetation for agriculture.

Plant communities (the various plants growing together in a particular habitat), therefore, represent the interaction of physical, biotic, and anthropological factors. **Physical** factors include rock type, climate, and soil. Consequently, they are sometimes subdivided into lithospheric, atmospheric, and edaphic categories. **Biotic** factors involve the interaction of plants and animals. **Anthropogenic** factors are human activities.

The orderly progression of a plant community from, say, the initial colonisation of a new bare rock habitat to the climax vegetation for that particular environment involves **plant succession**. The sequence of plant communities at any one site, through which this climax community is eventually attained, is known as a **sere**. In '*primary succession*' (a **prisere**) the first community is called the **pioneer colony**. The succession starts here with a **ground layer** of lichens and mosses because they can grow without soil. As weathering subsequently breaks down the rock into its constituent minerals, and dead plant remains are processed into humus by bacteria, the **field layer** of herbs and grasses begin to grow. These are taller than the ground layer, so become dominant until fast-growing shrubs take over. Taller plants will always dominate smaller ones by effectively taking light for photosynthesis from lower down. Consequently, the **shrub**

layer will ultimately be taken over by a **tree layer** of taller, slower growing trees. Providing that the physical, biotic, and anthropogenic factors do not alter, the vegetation should reach a state of equilibrium at this point. Since climate is likely to be the main controlling factor, this **climax community** is sometimes referred to as the '*climatic climax vegetation*.' However, human interference frequently alters the natural vegetation - hence **plagioclimax communities** are more likely. Indeed, the distinction between the prisere described above and '*secondary succession*' (a **subsere** whereby the plant succession progresses in an area already colonised) is often, although not always, a result of past human activity - such as the regeneration of an abandoned forest clearing. Sometimes a community within the plant succession, known as a **seral stage**, will be unable to progress to true climax vegetation because of repeated interference by outside factors, such as sea spray maintaining marram grasses on a sand dune. Finally, we distinguish major climax communities of flora (plants) and fauna (animals) corresponding to climatic regions as **biomes**. For example, the tropical rainforest (selva) and boreal forest (taiga) are both evergreen, yet provide dramatic contrasts in almost every other respect. The selva invites superlatives unmatched by any other environment. Beautiful, unique, and rich in biodiversity, the taiga appears impoverished by comparison.

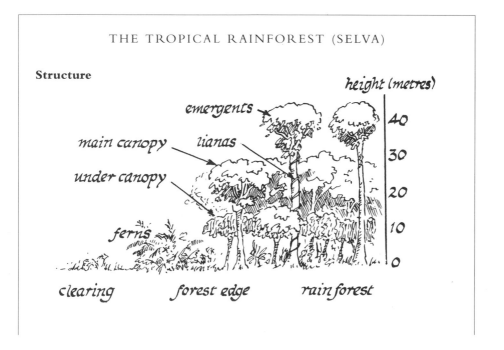

THE TROPICAL RAINFOREST (SELVA)

Structure

height (metres)

emergents
main canopy   lianas
under canopy
40
30
20
ferns
10
0
clearing   forest edge   rainforest

Flora is stratified into five layers (three tree, one shrub, and one field). Discontinuous emergents protrude 5-10 m above the main canopy which marks the general level of the rain forest. The main canopy consists of tall trees (c. 30 m high) with long slender trunks and rounded crowns. Beneath these an under canopy of shorter trees (c. 9-15 m high), with narrow crowns, is likely to be discontinuous unless there are breaks in the canopy. The shrub layer will be sparse due to lack of light. Even sparser will be the field layer of herbs, ferns, and seedlings, some growing to true ferns. Lianas (woody creepers) and epiphytes growing attached to the branches of other trees are common. Many of the canopy trees and emergents will have plank buttresses to support them. The bark is smooth and thin because there is no need for protection from the cold or fire in natural circumstances. The leaves are dark green, leathery to keep rigidity without wilting, shiny, and with **drip tips** to allow easy shedding of water. Almost continuous photosynthesis and transpiration is therefore assured. Indeed, the rainforest is characterised by simultaneous flowers and fruit, germination, regrowth, and leaf shed. There is no annual rhythm, therefore, and up to 200 plant species per hectare are supported.

Fauna, likewise, is diverse and abundant. 90 per cent of wildlife species are supported - all relating to the **stratification** (layering) of plants which creates numerous **ecological niches** with different foods, concealment, and movement opportunities. The canopy, for example, has monkeys, flying squirrels, butterflies - even frogs, which never go lower. Tree trunks are home to baboons and various insects whilst the ground will have termites, ants, and ant-eaters.

Soils are particularly interesting because of their paradoxical infertility. The high temperatures and humidity allows rapid weathering and so release of minerals, but also rapid leaching. Hence the acidic, tropical red (ferruginous) soil, dominated by clay, may be up to 45 m deep - yet with a very thin (c. 3 cm) organic layer.

Functioning

The nutrient cycle of the soil is confined to the organic layer and is almost closed and **leak free**. Materials are, therefore, recycled very quickly allowing little time for them to be lost through leaching. The closed cycle of growth and decay make the fertility dependent on the plants. Decomposers (bacteria and fungi) work very quickly in the heat and moisture - the nutrients released subsequently absorbed by the roots within the top 5 m. However, due to the excessively deep soil profile, minerals released from bedrock weathering have no chances to reach the plants. Consequently, should the natural vegetation ever be cleared, the soil is vulnerable and liable to rapid degradation.

Human interference

For generations, rainforests in South America, Africa, and Asia have supported shifting cultivators in relative harmony. Small plots cleared, cultivated, and then abandoned allowed (secondary) plant succession from weeds and grasses, to shrubs and seedlings, to canopy species of climax vegetation, representing full recovery, within 20-30 years. However, contemporary population pressure is not allowing sufficient time for this, with plots returned to too quickly. Consequently, the fresh clearance and burning kills seedlings and the succession is **arrested**. Soil and litter stores may initially be larger, but in time the soil leaches, leading to permanent deterioration. However, even more alarming is the annual destruction of an area equivalent to England and Wales by felling, bulldozing, and burning for commercial logging, mining, ranching, cash cropping, coffee, banana, and rubber plantations. The United Nations Food and Agricultural Organisation (FAO) estimates that there may be no rainforests left in 50 years if the accelerating rate of loss is not checked. Whilst restraint is needed before predicting global catastrophe, there can be no doubt as to the environmental threats such destruction poses. Soil fertility is reduced by leaching - with the danger of erosion by sheetwash or aeolian action (without protective vegetation cover). Indeed, without soil the vegetation cannot regenerate which may lead to desertification as already demonstrated in Brazil. The aridity this implies is promoted by lower precipitation resulting from reduced transpiration. There is, therefore, the spectre, however obscure, of a loss of climatic control. This cannot be ignored. For example, the **mirror effect** of dark-coloured, heat-absorbing forest cover undoubtedly has an influence on climate. However, this may prove to be only fully understood in retrospect, given that light-coloured, heat-reflecting deserts are known to change atmospheric circulation. More certain is exacerbation of the **greenhouse effect**, with less forest to absorb carbon dioxide in photosynthesis and, perversely, burning adding to the heat-absorbing gases.[1] Finally, the loss of **biodiversity**, with unknown commercial potential in the form of new foods, crops, oils, and medicines is frightening. 40 per cent of all known drugs originated in the forest environment - for example, snakeroot used to control high blood pressure, rosy periwinkle treating leukaemia, quinine for malaria, and not least yams in contraceptive pills. The 1992 Earth Summit in Rio de Janeiro failed to agree a treaty to save the world's rainforests, as ELDCs feared attempts to hinder their development. Future generations may rue this missed opportunity in a multitude of ways.

[1] Carbon dioxide and other pollutants emitted by the burning of fossil fuels since the Industrial Revolution act just like a glass greenhouse. They allow short wave radiation from the sun through to the earth, but then trap some of the longer wavelength radiation that would otherwise be reflected. Consequently the world gets hotter - changing climatic interrelationships and sea levels (through thermal expansion and ice cap melting).

THE BOREAL FOREST (TAIGA)

Structure

Flora at climatic climax is restricted to one or two species of coniferous trees. The cause of this limited vegetation is debatable. The lack of time since deglaciation is probably of great relevance, although fire is sometimes suggested as a factor. The conifers are characteristically tall (c. 25 m) with straight trunks. They form a canopy beneath which bare branches die due to shortage of light. Beneath these, the shrub and field layers are almost non-existent, again because of shortage of light, but also due to the hostile conditions for growth. The podsol soils are thin, acid, and infertile. Topped with a thick layer of resinous, acidic needle litter, seedlings simply cannot grow.

The harmonious relationship between vegetation and climate is worth stressing. Conifers replace deciduous trees when the growing season is less than six months and where less than four months are frost-free. (Trees cannot grow when temperatures fall below 6°C.) It is notable that the northern limit for conifers is set by problems of seed production and germination, not by conditions for mature trees to survive. Indeed, the conifers have numerous characteristics evolved to cope with particularly harsh environmental conditions:

♦ The trees cannot take up water at low temperatures or when the ground is frozen. Consequently, in order to survive they must stop or at least reduce transpiration. This is achieved by the stomata closing in winter and by the thick cuticle and small surface area of the needles. Thick bark also reduces transpiration losses.

♦ The conical shape of the trees along with flexible, drooping branches and thin, waxy needles allows snow to be easily shed. This prevents damage in winter.

♦ The dark leaf colour helps the trees absorb solar radiation for efficient photosynthesis during the short summer season.

♦ The shallow root system allows the trees to absorb water as soon as the ground surface thaws.

♦ The trees are acid tolerant and so thrive in podsols. However, it is a symbiotic (two-way) relationship in that they partly cause podsolisation.

♦ The evergreen nature of the trees allows them to carry out photosynthesis and transpiration as soon as the warmer weather starts.

Fauna is clearly restricted by the sparsity of habitats. Deer in clearings, for example, will browse on seedlings whilst squirrels, martens, and birds eat cones. The acidic forest floor limits insects, but they may live on the bark. Finally, burrowing animals are restricted by the permafrost.

Soils are particularly distinctive podsols, characterised by severe leaching during the spring thaw. Sequioxides of Fe_2O_3 and Al_2O_3 are washed out of the zone of eluviation in a process known as **chelation**. This leaves a bleached A horizon beneath which iron and aluminium accumulate in the zone of illuviation. This B horizon is coloured red and yellow in consequence As previously discussed in *Soils*, podsols develop best where acid leaf litter occurs on freely drained material such as sand, and where precipitation exceeds evapotranspiration.

Functioning

As with all forests there is a large biomass store. However, the boreal forest is associated with a particularly large litter store due to limited decomposition. This is because of low temperatures in the summer shade, winter frost, and the resinous needles being difficult to break down. Litter, consequently, accumulates on the forest floor, slowly breaking down into acidic (mor) humus. The store of nutrients within the soil, therefore, is very small and nutrient recycling particularly slow.

P.F. Gersmehl's (1976) **model of the mineral nutrient cycle** demonstrates the differences between ecosystems in terms of nutrients stored within the biomass, soil, and litter. The contrasts between the selva and taiga biomes are particularly well illustrated using this system.

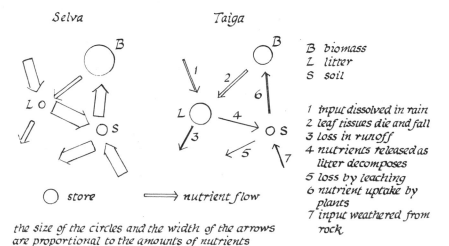

Selva   Taiga

B biomass
L litter
S soil

1 input dissolved in rain
2 leaf tissues die and fall
3 loss in runoff
4 nutrients released as
   litter decomposes
5 loss by leaching
6 nutrient uptake by
   plants
7 input weathered from
   rock

○ store   ⟹ nutrient flow

the size of the circles and the width of the arrows
are proportional to the amounts of nutrients
stored and transferred respectively

So far the ecosystems and biomes studied have included references to plant succession, but mainly at the global scale. A study of the prairie grassland ecosystem would be an appropriate regional scale example. It is only at a local scale, however, that the detailed plant relationships within priseres and subseres may be fully appreciated.

For example, a **lithosere** occurs when bare rock is progressively colonised as soils develop. The volcanic island of Surtsey has proved particularly revealing in this respect, because scientific monitoring has been continuous since the original eruption in late 1963. However, retreating glaciers provide more accessible research opportunities. W.S. Cooper working in Glacier Bay, Alaska, from 1916 to 1931 observed succession from pioneer to climax in only fifteen years. It was notable that the plants themselves changed the habitat. It was, therefore, an **autogenic succession** starting with moss, willow herb, and horsetail as the pioneers helping soil formation. The shrub stage saw a ground covering of low growing shrubs and thickets of dwarf creeping willows and increasingly dominant alder shading out the herbs. Finally by the tree stage, the alder had improved the nitrogen content of the soil so much that sitka spruce could become the climax vegetation. Similar studies, today, are possible in the Alps.

Hydroseres associated with a fresh water pond or lake might show a succession of submerged and floating plants, such as lilies, with reeds and rushes encroaching at the water's edge and encouraging sedimentation. Willow and alder with ash and oak climax would be the likely progression, taking over the whole site once the pond silts up entirely.

Haloseres are best demonstrated in the inter-tidal salt marshes of estuaries and sheltered bays. The mudflats are exposed for short periods twice a day. Consequently, the pioneer vegetation must be tolerant of inundation, salinity, and the mobility of the environment. Halophytes, such as spartina grass, thus become established in this so-called **slob zone** - checking the water flow and promoting silting. This raises the surface level with, consequently, less time inundated. The resulting **sward zone** habitat is, therefore, less salty and so tolerant to non-halophytic species. Once the level has risen to the point where only high spring tides cover it, sea holly and blackthorn may colonise, progressing further inland to a climax of oak and ash.

Psammoseres are best illustrated along transects across coastal sand dunes. Lyme grass, sea couch, and sea twitch invade embryo dunes – the latter with marram grass helping stabilise the material. Indeed, marram with its long roots and shiny, folding leaves to reduce evapotranspiration, is ideally suited to the exposed, arid dune environment and so likely to dominate the main (**yellow**) ridges. Further inland, older (**grey**) dunes are more likely to be fixed by mosses, lichens, and herbs with the establishment of shrubs such as hawthorn, gorse, and broom leading to a climax of oak or beech.

However, the local scale is increasingly likely to reveal plagioclimax communities whereby human interference permanently arrests the succession at a seral stage – preventing the natural climatic climax vegetation from ever being established.

THE NORTH YORKSHIRE MOORS

The plagioclimax vegetation of this area is permanently maintained by **muirburn**. Muirburn is the controlled burning of heather (*calluna vulgaris*) in order to produce and maintain an ecosystem that would not otherwise have existed. Heather, and many of its associated species, is a valuable evergreen forage plant. It is the staple diet of red grouse and a major food source for hardy hill sheep - hence the commercial interest in its maintenance.

Heather is a woody shrub, tolerant of a wide range of climate and soil. Under natural conditions it can grow at a maximum altitude of 500-600 m. Originally it was probably a widespread, but minor, element of forest and woodland on acid soils. It is unlikely to have dominated unless exposure or extreme soil conditions inhibited tree growth. The spread of heather-dominated areas started, therefore, with the clearance of forests and was aided by increasing numbers of domestic grazing animals (cattle and sheep) which prevented the regeneration of seedlings.

C.H. Gimmingham (1975) recognises four stages of growth:

1. The **pioneer stage** is the first 6-10 years when heather seedlings are establishing themselves and the root system is growing more rapidly than the shoots. However, at this stage the shoots do have a high nutrient content.

2. The **building stage** lasts for the next 6 years. This is the most productive and valuable stage, characterised by vigorous branched growth, prolific flowering, and an increase in biomass to form continuous ground cover.

3. The **mature stage** lasts for the next 9 years during which the ground cover becomes discontinuous and the plants woodier, with a recumbent form.

4. The **senile/degenerate stage** sees the growth of plants older than 25 years slowing down. Their maximum height is reached and the oldest central branches die off. At this stage large gaps in the ground cover are notable and the value of the heather is very limited.

The object of muirburn is to keep as much of the moorland in the building stage as possible. The surface heather is, therefore, burnt off in a 10-15 year rotation dependent on site conditions. This will keep as much heather as possible at its most productive stage, with the proportion of edible green shoots at its highest. The total area burnt in any one year will depend upon the length of rotation. It is usually done in small areas of about one hectare. This not only allows the firing to be carefully controlled but also provides a feeding area for one pair of breeding grouse with unburnt nesting cover nearby. Normally there are about six one-hectare burning patches within each square kilometre resulting in a 'patchwork quilt' pattern of heather at various stages of regrowth. Repeated burning eventually produces the plagioclimax heather monoculture.

The effects of muirburn are varied. Old woody plants are destroyed - releasing their nutrients in ashes available to support new shoots. Clearly great skill is required in assessing when to burn, for the plants must be at the end of the building stage, but not too mature or senile or the regrowth will not take place. Also the burning must not be too severe because new shoots develop from buds on root stocks just below the ground surface. Normally protected from fire by heather litter, excessive burning would destroy them. After burning the soil pH rises, because nutrients are released as described earlier and, as the peat burns, minerals are drawn from the B horizon to the surface. However, leaching rapidly reduces it again. Also some species, such as bilberry, grow faster than heather and so may establish themselves. However, just as the raised pH was temporary, so is this, with the heather re-establishing itself and choking out the invaders to regain dominance.

The typical soil under heather moor is a peaty gleyed podsol. This is because of the cool, wet climate whereby precipitation exceeds evapotranspiration. Acid parent rocks promote this too, yet calcareous grits common on the North Yorkshire Moors do not counter the acidity because the calcium is dissolved and leached downwards.

Heather is suited to this acidic soil because it has thick cuticles on needle leaves with small stomata. This reduces evapotranspiration and helps it resist the dehydrating effect of strong winds. It also prevents fatal toxins of iron, manganese, and aluminium building up in any great concentration. Indeed, *calluna vulgaris* produces organic compounds which fix the iron and aluminium in a harmless state, so preventing damage to the plant. However, when the heather is burnt or dies these are released back into the soil.

Finally, the soils have not always been peaty gleyed podsols. They reflect the contemporary plagioclimax conditions. Three thousand years ago, however, the area was covered in deciduous woodland on fertile brown earth soils. Forest clearance, the start of burning, and the consequent vulnerability to leaching resulted in the podsolisation process. It is a valuable reminder that human interference will inevitably have environmental consequences.

10

POPULATION

Population geography should be distinguished from **demography** (the science of population) by its environmental context and emphasis on **how** and **why** demographic patterns and processes change spatially (from place to place) and temporally (over time).

The United Nations Organisation (UN) predict a world population of over 10 billion by the year 2100. However, the rate of increase in ELDCs so exceeds that in EMDCs that 86 per cent of this total is predicted to be in the so-called Third World. Given that contemporary overpopulation in many regions is already evident, plus millions of displaced people through conflict, hunger, and poverty, it is clear that geography has an important contribution to make to understanding and planning for such realities. However, we can only be precise in describing, classifying, comparing, and analysing population patterns with accurate data sources.

Population data sources

There are three main sources - the census, registers of population, and global surveys. The **census** is a glorified 'head count' on a stated day, asking questions about household composition, ages, birthplace, occupation, workplace, education, housing arrangements, amenities, and so on. The first British census was held in 1801 and, excepting 1941, has been held every ten years since. It takes a *de facto* approach by indicating where people are residing on the actual day of enumeration (usually the third Sunday in April). A contrasting method is the *de jure* approach which records individuals according to their normal place of residence. This method is used in the USA.

Generally speaking, good quality, reliable statistics are available in EMDCs such as Britain and the USA. However, there are always problems such as whether or not temporary workers abroad should be included. Likewise, should sailors at sea be excluded? Where do you record as resident a worker employed in London but staying with his family in Hull at the weekend? How (and should) you question on sensitive issues such as ethnic group? It is of note that this particular 'nettle' was finally grasped in the 1991 UK census by means of the same inoffensive tick boxes used in much of the rest of the questionnaire - with none of the 'political fall-out' anticipated by many civil rights campaigners. However, these problems pale into insignificance when compared to the problems of enumeration in ELDCs.

COUNTING HEADS IN ELDCs

World census data is of variable quality. Most EMDCs are able to produce regular, reliable counts of their populations and hence valuable data for their governments and social planners. Such counts, however, are expensive and require great planning. It is in ELDCs in particular where enumeration is difficult. Whilst successful in some, such as Kenya, in many such countries, the census is either inadequate or non-existent. There are several important reasons for this:

1. Poor and inadequately financed methods of data collection are commonplace. ELDCs are often short of money for necessities, such as food, and have little to spare for the census. Enumerators are usually essential to collect the data as levels of literacy are low – limiting the number of people who could accurately complete their own census form. Hence, training of enumerators is vital for the census, but in itself is expensive.

2. Population mobility is frequently a problem. In many tropical countries, such as Nigeria, many people are nomadic, hence easily omitted from the count, or included repeatedly! Movements into and within the cities of ELDCs can be especially high. It is difficult to ensure, therefore, that everybody is included in the count. Population patterns also change rapidly with new shanty towns, for example, forming.

3. Owing to patterns of employment, men who migrate to the cities, or nearby countries, for work may have several residences. Should they be included in, say, the city count where they work or are they really rural residents where their families remain?

4. The level of homelessness in some cities, such as Calcutta, India, is so high that it is difficult to ensure that all street dwellers are included. Knowledge about this group of people, because of lack of shelter, work or whatever, is clearly of importance when planning in ELDCs.

5. People are often suspicious of the questions asked and have no idea of how accurate information may directly help them. False statements, especially about age and occupation, are common.

Data may not be comparable for neighbouring countries as the counts may be conducted at different dates. Also terms within a census may not be clearly defined. For example: What is your mother tongue? How many are in your household? How many rooms are in your house? All of these questions are open to varied interpretations by different people.

> In recent years the UN has assisted many ELDCs with both money and expertise to
> ensure that census data collected is worthwhile. Even so some countries have not
> recently held a census.

Registers of population are records of births, marriages, deaths, and
divorces. In the UK, for example, it is illegal to withhold notification of
such events within set time periods. The statistics are subsequently collated
by the General Register Office. The Registrar General then publishes
quarterly and annual updates which, whilst far more accurate (albeit
limited) than the census, are not immune to double entries or omissions.

Global surveys have become increasingly important. For example, the
World Fertility Survey (WFS) overseen by the UN between 1974 and 1984
proved to be one of the most important social studies ever undertaken,
providing a wealth of information to governments formulating national
population, economic, and social policies.

A concluding note on population data sources should, however, emphasise
the most difficult problem faced by the population geographer. Despite
UN efforts at standardisation in census procedures, different categories of
questions, classification systems, and enumeration years and intervals make
comparing different countries very difficult. Also, many countries only
held their first census recently, making historical comparisons impossible.
Finally, even now, over one-quarter of the world's people escapes regular
enumeration, hence the importance of multinational efforts such as the
WFS.

Population distribution

The distinction between population distribution and population density
must be understood. **Population distribution** is simply where people
are located. **Population density** expresses the ratio between total
population numbers and the area occupied by that population.

Crude population density (average number of people per unit area) is of
limited value because of the uneven distribution likely within enumeration
districts. Therefore modifications such as nutritional density, occupational
density, and room density are sometimes quoted.

Some geographers still refer to ecumene and non-ecumene. **Ecumene** is
the inhabited areas of world – approximately 60 per cent of the land

surface. **Non-ecumene** is the uninhabited or very sparsely populated remainder.

However, this twofold distinction is open to criticism primarily because of the difficulty in delimiting the graduated zone between the two - known as the **frontier of settlement**. A threefold division into high, moderate, and low density with regard to comparative standards of living is, therefore, more useful.

MAPPING POPULATION DENSITY AND DISTRIBUTION

Population density is usually shown cartographically by means of **choropleth** maps. These use a series of shadings to show density values. The effect, however, is of a uniformity which suddenly changes at a political boundary. Indeed, the fundamental weakness of all density maps is that population totals are related to units of area within which, in real life, there will not be a uniform distribution of population. Information, therefore, is obscured by the average. For example, the most striking examples of population concentrations occur in urban areas. This will be obscured, however, if averaged with the surrounding countryside.

To use smaller units of area would be more accurate, but it is difficult to obtain data. Also, the break line between areas may not be as clear cut as the map implies. And what divisions should be used? Physical regions, administrative units, grid squares? Further misunderstanding can arise if the division of density values for the key is carelessly done.

For example: density class 21-40 per km². 19 and 22 per km² fall into separate categories even though their values are much more similar than, say, 21 or 38 per km², which fall into the same density class.

People per sq. km

161 and over
81 - 160
41 - 80
21 - 40
0 - 20

Population distribution tends to be mapped using **dots** and/or **proportional circles (spheres)**. The dot map allows pin-pointing of the location of population as precisely as possible. For example, the problem of showing urban population concentration in an area may to some extent be solved by dot maps - which are the most common method of mapping population distribution.

The choice of dot value is critical. Good practice is to choose one which will have at least a few in the lowest category and not too many in the highest.

The size of the dot should be small enough to allow for as much flexibility as possible, whilst remaining clear enough so as not to be confused with a speck on the paper. Dots that are too big tend to overlap, which is undesirable. Hence problems include the decision of how many people per dot and the physical size of the dot.

But where should the dots be placed? Their precise location will be a difficult choice, not least because they suggest so much. Also, it is often impossible when using dots with a low value (often needed for sparse rural areas) to effectively show large towns. This is when a choice of, say, two dot sizes or proportional circles may be appropriate.

If dots are spread evenly over an area a uniformity that may be misleading is suggested. This then suddenly changes at an administrative boundary - a similar problem to the density map. By the time one then starts questioning issues such as reliability of data, nomadic people, and so on, the difficulties of the cartographer's task start to be appreciated. Any map, therefore, is a compromise and, clearly, instantly dated.

Should it be necessary to measure the extent to which the population is concentrated around a central city or clustered in a particular district, or uniformly spread or randomly scattered - **nearest neighbour analysis** (explained in *Settlement*) may allow the problem to be quantified. However, this objective technique would involve measuring the subjective dot positioning of the cartographer and so must be adopted with a degree of healthy scepticism.

The extent to which population distribution is concentrated can be plotted in graph form by means of the Lorenz curve. This is particularly useful in studying changes in population distribution through time.

Lorenz curve

For the territory to be studied, such as a country, the areas and population totals of sub-units, such as counties, must be known. The population and area of each county is worked out as a percentage of the total. Then they are arranged in descending order according to their population totals. Cumulative percentage of population is subsequently plotted against cumulative percentage of area. The points are then joined to form the Lorenz curve.

Changes in the form of the curve through time show whether population distribution is becoming more concentrated, dispersed, or remaining static. A totally dispersed population is represented by the diagonal line. The nearer the curve to this diagonal, the less concentrated is the population distribution. Therefore, the nearer the curve is to the *x* axis, the more concentrated the population becomes.

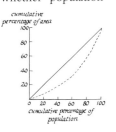

80 per cent of the world's population live on 20 per cent of the land surface. 90 per cent live in the northern hemisphere and 50 per cent in South East Asia alone. It is clear, therefore, that population is not evenly distributed throughout the land masses. Some areas are densely populated – others almost totally uninhabited. The reasons for this are legion, but one is of ultimate importance - **economic potential**. People will only live where they can find a means of earning a living, even if only at subsistence levels. Factors allowing this can be physical, economic, social, and even political.

One must not be tempted into a superficial examination of distribution by concluding that physical influences, such as climate and terrain, solely determine the pattern. This type of interpretation, where patterns of settlement and economic activity are held to be determined by environmental factors alone, is a view known as **geographical determinism** and was stated by E. Huntington and others earlier this century. It is now realised that humans are capable of modifying and controlling the environment to a considerable degree – exploiting environmental possibilities and opportunities to the full. We are the dominant element in the relationship between people and the environment – a viewpoint referred to as **geographical possibilism**.

Although modern technology enables us to live on almost any part of the earth's surface, certain areas have tended to discourage settlement. This is in part due to physical discomfort, but also because of low economic potential, dictating few opportunities for us to find a livelihood. Such areas include zones of extreme aridity, high mountains, and regions with very low temperatures. Other parts of the world, such as the agriculturally productive alluvial plains of southern Asia, offer particular physical attractions – hence the dense populations.

It is of note just how predominant physical factors influencing population distribution are. Indeed, one could argue that **accessibility** and **water supply** should head any list. However, reflection notably concentrates on agriculture as a theme, because around half of the world's people still live directly by farming, and of course food production is of paramount importance to us all. Consequently, areas where agriculture is impossible, difficult, or expensive to undertake will be sparsely settled – for example, mountainous areas with poor soils or where the climate is too hot, too cold, too wet, or too dry. It follows that low-lying areas, especially in the temperate zone, are particularly well populated.

INFLUENCES ON POPULATION DISTRIBUTION

1. **Relief:** Steep gradients and rugged terrain restrict movement and deter settlement. Lowland plains with gentle or flat relief, however, encourage agriculture and high population densities.

2. **Altitude:** There are few permanent settlements above 5000 m because the low temperatures and thinner atmosphere restricts comfortable habitation. Consequently, 80 per cent of the world's population lives below 500 m – with 56 per cent between sea level and 200 m.

3. **Climate** and **weather:** Extremes of cold and aridity, such as found in hot and cold deserts, deter settlement. It is argued that temperate climates, such as in western Europe, provide ideal conditions – yet as stated earlier, half the world's population live in South-East Asia which has a monsoon climate.

4. **Soils:** Thin, infertile, or badly drained soils deter agriculture and so settlement. Fertile deltaic and alluvial soils, by contrast, encourage agriculture.

5. **Natural vegetation:** Particularly in the past, should this be difficult to clear and/or result in low agricultural potential, such as tropical rainforests, settlement would be deterred. Conversely, areas of high agricultural potential after clearing, such as the temperate deciduous forests of western Europe, are more densely populated.

6. **Mineral** and **energy resources:** Few or no resources deter settlement just as readily available reserves encourage it. For example, the distribution of western Europe's population is closely associated with coalfields.

7. **Natural hazards:** All logic dictates that hazardous areas would be avoided – but this is not always the case. Paradoxes abound whereby fatalistic attitudes overcome fear for sake of associated advantages. For example, the highly fertile volcanic areas of Java, Indonesia, and the flood plains of Bangladesh support some of the densest rural population concentrations on earth.

8. **Economic structure:** Levels of industrialisation, urbanisation, and technology are relevant in that generally the more developed and sophisticated the economy, the more people can be supported. Urban industrial areas by definition are densely populated.

9. **Culture, tradition, religion,** and **politics:** Arguably peripheral influences, yet not without significance in specific cases.

Continuing the agriculture theme, perishable food cannot usually travel great distances, hence large-scale farming must take place in areas with easy accessibility, such as along valleys and near coastlines rather than too far inland. Only relatively recently, with refrigeration and efficient transport, have inaccessible regions such as the Prairie Provinces of Canada been opened up for food production.

Exceptions to the 'pull' of farming where population has grown up, is on or near mineral and energy resources. Generally, the more valuable the reserves, the more likely we are to overcome harsh natural conditions in order to exploit them. Hence, for example, silver and tin is mined at very high altitudes in the Andes and high grade iron ore in arctic Sweden (see *Minerals and Energy*).

Human influences on population distribution are inextricably linked to the physical, but certain specifics should perhaps be noted - not least the role of governments in economic and settlement planning and especially the age and duration of occupation in a given area.

Religion accounts for settlement in some areas. For example, it can be argued that the seventeenth century emigration of the Pilgrim Fathers from Britain and their subsequent colonisation of New England helped the American east coast to develop a high density of population. Certainly the proximity and nearness to the *Old World* of Europe, Africa, and Asia must be noted. Similarly, the nineteenth century settlement of Utah and its capital Salt Lake City was due to the arrival there of American Mormons.

Political factors have certainly created new areas of population. For example, in the decades following the 1917 Revolution, millions of Russians were forced to develop the wastes of Siberia. Also, one could argue that South-East Australia's emergence as the most populated region of the entire continent was due to its original development as a British penal colony - agriculture and industry consequently having had longer to establish itself.

Certainly it should be remembered that population distribution in any given area is the product of many influences - some physical, some human. It may therefore be difficult to determine the importance of any single factor or set of factors. Furthermore, the problem of analysis is made even more difficult because patterns of population distribution are always changing in response to changes in birth and death rates and migration trends. The world is not a static entity but constantly changes through time. Indeed, factors important last century may be insignificant now, as new influences related to economic and technological developments and planning decisions come into play. J. Beaujeu-Garnier (1978) expressed this admirably (in a quotation not devalued through time) by stating that '...*the general human cover is thickening over the globe...but it is thickening particularly in certain areas and becoming thinner in others...The population map for the beginning of the nineteenth century bears no resemblance to the present one and the latter probably looks very different from the one which future generations will see after the year 2000. However, through all the changing circumstances one can always see three great and fundamental influences: natural conditions, economic conditions, and the events of history.*'

POPULATION DISTRIBUTION IN AUSTRALIA

Australia's population distribution is peripheral, concentrated around the edges and mainly outside the tropics. The eastern coastlands are more heavily populated than the west. This is a result of the interaction of physical and human factors.

Physical: Climate is of primary importance. Latitude dictates the central tropical desert which discourages population. The indigenous people are small isolated bands of Aborigines increasingly relocated within urban centres and marginalised as a persecuted minority prone to alcoholism. Typically the desert stretches to the western coastlands, but on its eastern flank is bordered by a region of temperate grassland. Settlement is, consequently, more extensive and denser to the east where the cooler, temperate climate with adequate rain is notable.

Human: The northern coastlands are largely empty as a result of human rather than physical factors. Australia has operated a 'whites only' immigration policy which has meant that Asian immigration has been largely precluded. 'White' settlers have been unable and unwilling to develop the humid monsoonal northern belt which might have supported a dense Asian population. Economic factors have played a particularly significant role. Settlement of the interior areas of western and north-eastern Australia, for example, results from the exploitation of rich and varied mineral resources. The peripheral distribution is partly, also, a reflection of former colonial ties with Britain. Ports were the main points of exchange for exported food, such as wheat, and industrial raw materials, such as wool, and the import of manufactured goods. The settlers originated mainly in western and central Europe. The majority are urban dwellers and are attracted to the cities. Indeed, a striking feature of Australia's population geography is the dominance of the major cities in and around which settlement densities are as high as in countries with far larger total populations. Last but not least, Australia's distribution and density of population in part reflects the comparatively small total population. One could argue that many areas are sparsely populated because of relatively few people (c. 16 million) to settle a vast total area – 30 times larger than Britain.

World population growth

Today's great problem is not so much the size of the world's population but the rate of growth in those economically less developed areas least able to cope. UN global projections of replacement-level fertility being reached between 2010 (representing world population stabilising at 7.7 billion by 2060) and 2065 (with stabilisation at 14.2 billion by 2100) are dependent on the success or otherwise of ever-increasing numbers of national population policies. However, projections cannot make assumptions about future wars, economic systems, food supplies, climatic and environmental change, or whatever. For example, nuclear war seems increasingly less probable, but the impact of AIDS is still uncertain. Consequently, certainties in global predictions are a nonsense.

The origins of humanity are also the subject of considerable dispute although we can be traced back for over 1 million years. In these early times the numbers of people can be assumed to be very small. Indeed it was 1830 before the first billion was reached. However, the second by 1927, the third by 1960, the fourth by 1974, and the fifth by 1987 illustrates, dramatically, the staggering rate of growth seen by many as a serious threat to our survival. Indeed, the use of phrases like **population explosion** tends to encourage gloomy predictions – especially given that it is not evenly spread. In EMDCs growth is sluggish to the point of non-existence, whilst in ELDCs rapid growth, poverty, and hunger mutually perpetuate each other.

If this situation is to be understood, however, it is important to examine what has happened in the recent past in order that likely future developments in specified areas can be more fully appreciated and assessed.

In earlier times humankind gathered and hunted or else lived solely by agriculture. Even in regions now considered developed, existence was 'hand to mouth' and dictated by the will of nature. Only the fittest survived regular population checks of plagues, diseases, famines, floods, droughts, fires, tribal wars, and so on.

Today, few parts of the world remain like this. Indeed, the last 400 years have seen fundamental changes. Britain, for example, experienced so-called Agricultural and Industrial Revolutions which were to have untold effects on economic, technological, and scientific progress the world over. On the one hand, increased economic productivity could support more and more people – on the other hand improvements in medicine and hygiene, flood control, and fire precautions, for example, lessened the effects of natural population checks. The result was that more children were being born and, since the chances of survival were increased, fewer people were dying.

As long as birth rates remain higher than death rates, population grows. The wider the gap, the greater the growth. Yet what are the many and complex factors which determine fertility and mortality? Both are the outcome of a combination of economic, religious, social, cultural, and political considerations.

COMPONENTS OF POPULATION CHANGE
(P. Haggett, 1975)

In any area there are two basic ways in which population change takes place:

1. **Natural change** as a result of people being born or dying.
2. **Migration change** as a result of people moving in or away.

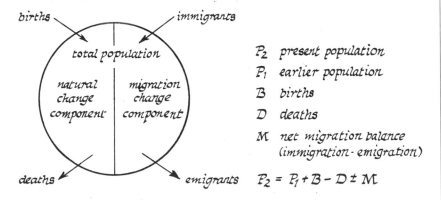

P_2 present population
P_1 earlier population
B births
D deaths
M net migration balance (immigration - emigration)

$$P_2 = P_1 + B - D \pm M$$

Total population represents the balance between the natural change and migration change components, the relative significance of which will vary greatly from place to place – both spatially and temporally. Clearly only the natural change component need be considered when dealing with the world as a whole.

Natural increase (NI) = crude birth rate (BR) - crude death rate (DR)
(normally expressed
as a percentage)

$$\text{Population doubling time} = \frac{70}{\text{per cent growth rate}}$$

Fertility

The **birth rate** is the most common index of fertility. This is the ratio (expressed per thousand) between the number of live births in one year and the total mid-year population.

$$\text{Crude birth rate} = \frac{\text{total no. of live births in 1 year}}{\text{total mid-year population}} \times 1000$$

It is tempting to assume that birth rates are linked to economic advancement. Certainly fertility tends to decline where living standards improve but remain high where technology is backward. However, a plethora of other factors must be considered such as moral, intellectual and financial motives, attitudes towards marriage and children, the status of women in society, religious and superstitious beliefs – even the extent of material ambitions in modern living. All have a bearing on birth rates.

Demographic (age-sex) structure is fundamental in that high proportions of young adults, such as in New Towns, would promote higher birth rates than regions with a high percentage of children (**juvenility**) or aged. However, the former suggests high fertility to come.

Religions such as Catholicism and Islam engender complex attitudes to family planning, despite the former's strict teachings on artificial contraception. Certainly fertility is normally higher in countries dominated by these faiths. The contemporary paradox of Italy, however, as spiritual home of the Catholic faith, yet with the lowest recorded birth rate in the history of humankind is worth noting.

Education brings choice, greater social awareness and, not least, a knowledge of birth control. Consequently smaller families are associated with more advanced education, social standing and emancipation of career minded women.

Social and **cultural customs** have great influence. Hindu girls in India, for example, marrying at an early age encourages large families. Likewise **polygamy** increases fertility, as does the necessity for a male heir.

Health and **diet** is relevant in that high mortality may necessitate high fertility in order that some children survive into adulthood and the providing role.

Politics directly influence fertility by, for example, sanctioning the financial inducements, job, and pension protection for pregnant women in a pre-unification West Germany alarmed at its ageing population. Whether the high cost of unification will allow such policies to continue remains to be seen. This example is illustrative of the contrast in population policies between EMDCs and ELDCs. Wars also have notably immediate effects on fertility, with a fall in the birth rate during hostilities followed immediately by **fertility bulges (baby booms)** such as in Europe between 1918 and 1920 and 1945 and 1949.

Birth rates tend to be particularly prone to short-term fluctuations in EMDCs where the state of the economy can, with subtlety, influence them by decreasing fertility in times of recession and increasing it in times of optimism. Likewise, the vagaries of fashion can never be underestimated, or the impact of the media and even arguably trivial events such as procreating royalty or power cuts!

Mortality

Death rates tend to be more stable and predictable than birth rates. However, one must not underestimate its importance because it is the decline in mortality, rather than any increase in fertility, which is largely responsible for population growth.

$$\textbf{Crude death rate} = \frac{\textbf{total no. of deaths in 1 year}}{\textbf{total mid-year population}} \textbf{ x 1000}$$

Again we find a link with levels of economic development in that the lowest death rates tend to be associated with regions of high living standards. However, beware simplistic correlations – for one may find relatively high death rates in EMDCs, associated with **diseases of affluence**, compared to many ELDCs where 'western' influences in health care, hygiene, nutrition, sanitation, and so on have allowed 'death control' without the excesses of affluent societies.

Demographic structure is reflected in higher mortality in regions with ageing or male dominated populations. The former is very evident, for example, in Britain's south-western coastal 'retirement resorts'. The latter reflects marked female longevity.

Health care is critical in that the better the medical services and doctor–patient ratio, the lower the death rate.

Social 'class' has great relevance in that many poorer sections in any society suffer less privilege, sub-standard housing, imbalanced diet, and worse medical care.

Occupation and **abode** are related in their significance in that the stresses of urban life and responsible jobs may lead to nervous strain and early death. Less pollution, less crowding, and lower traffic densities in rural areas may increase longevity. However, certain manual jobs, especially in primary occupations such as mining, fishing, and farming, are not without their hazards too.

Infant mortality is the number of deaths of infants under one year old expressed per thousand live births. It is a particularly sensitive indicator given the vulnerability of this age group. For example, up to one-third of the total annual deaths in central Africa are of infants. Infant mortality rates vary enormously throughout the world, but tend to be highest in poor countries with high child mortality too. As with mortality in general, any decline benefits urban areas first.

Causes of death tend to be classified as either exogenetic or endogenetic. **Exogenetic (environmental)** causes include infectious diseases due to contaminated food or water, bad housing, and so on. **Endogenetic (degenerative)** causes include failure or gradual collapse of body functions through cancer, heart disease or senility.

Exogenetic causes dominate in countries with high death rates, whereas endogenetic causes are associated with lower death rates. For example, 75 per cent of all deaths in Britain are due to heart disease, cancer, or cerebrovascular disease – the so-called diseases of affluence.

In conclusion it should be noted that this century has seen a global reduction in death rates as a result of improved medical services and immunisation campaigns against infectious diseases. Infant mortality has been reduced and average life expectancy increased. Reduction of death rates means more mouths to feed and more aged people to support. Therefore, unless these reductions in death rates are matched by increases in resource development and widespread birth control, one is simply creating new problems.

THE DEMOGRAPHIC TRANSITION MODEL

The fact that birth rates and death rates tend to decline with economic and technological progress goes some way to explaining why EMDCs have considerably lower population growth rates than ELDCs. It also helps to explain why such growth rates in any given region may change over time. Europe, for example, over the last 1000 years has developed through a set pattern of change. This forms the basis for a general model of demographic transition from a primitive agrarian economy to a complex, modern, urban industrial one.

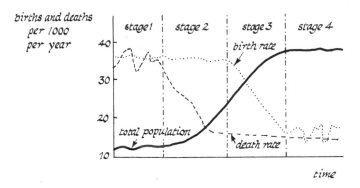

Stage one (high fluctuating, primitive regime) is characteristic of the economically least developed societies where high birth rates are accompanied by low life expectation and, consequently, high fluctuating death rates as population is checked by disease, war, and famines. Population growth is slow and intermittent.

Stage two (early expanding regime) is characteristic of many ELDCs today, whereby birth rates stay high and may increase marginally, but death rate declines progressively. The high birth rate reflects lack of birth control and sociological factors such as women marrying earlier. The falling death rate is related to economic growth and improvements in personal hygiene, sanitation, medical care, and diet. Population growth is rapid and accelerating as the gap between birth rate and death rate widens.

Stage three (late expanding regime) is characteristic of a more developed phase whereby the birth rate starts to fall as economic development and education weakens traditions and taboos, emancipates more women, and smaller families are desired and made possible by birth control. The lower death rate reflects the control of major diseases and improved standards of health and sanitation. Population growth continues, but at a progressively slower rate.

Stage four (low fluctuating, mature regime) is characteristic of highly developed societies where standards of living are high and both birth rate and death rate fluctuate around a low level. Fluctuations of the former may cause periods of actual population decline within the context of overall stability.

Criticisms of the model

The model is Eurocentric - yet countries in Africa, Asia, and Latin America have widely differing environments, racial, cultural, and historical backgrounds. To assume similar transitions, therefore, would seem unrealistic, especially given that ELDC base populations are higher, nor do they have any major outlets for the massive overseas migration which relieved population pressure during Europe's phase of rapid growth. Also foreign aid, and investment in agriculture, education, and family planning programmes are likely to shorten the time scales involved, providing entirely unique stage characteristics do not evolve regardless. Indeed, many would argue that a **fifth stage** is already required in order to illustrate contemporary demographic realities in many EMDCs. Whether this new stage should show population decline through birth rates falling for economic reasons, or death rates rising again as diseases of affluence and environmental problems in urban industrial societies increase, remains debatable. One should regard the model critically, therefore, with considerable reservation as to its general application. However, it undoubtedly provides a most useful conceptual framework for the study of population changes.

Demographic (age-sex) structure

This refers to the numbers of males and females of different ages within a given population. It is, in effect, a demographic history resulting from fertility, mortality, and migration factors which have operated during the lifetime of the oldest members.

Demographic structure is a factor of considerable social, economic, and political importance. For example, patterns of purchasing and consumption, the size and characteristics of the labour force, the type and range of welfare services needed for the population, and so on, will all be influenced by it.

The aged: Largely non-productive dependants containing a predominance of females.

Adults: Economically active and most mobile - occasionally subdivided into 'young' and 'older'.

Children: Largely non-productive dependants.

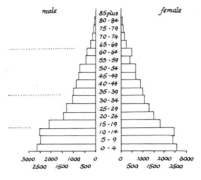

Demographic structure is most commonly shown diagrammatically using **age-sex (population) pyramids**. Long-term changes in fertility and mortality as well as lesser influences, such as persistent in- or out-migration, wars, and epidemics will be reflected in their shape. Model pyramids provide guidelines against which actual populations may be judged.

MODEL AGE-SEX PYRAMIDS

Stationary pyramids demonstrate stable fertility and mortality over a long period of time.

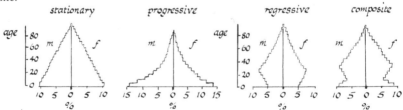

Progressive pyramids are bell-shaped and demonstrate high fertility and mortality characteristic of the poorest ELDCs. High juvenility, when added to the aged, determines **high dependency ratios** which make economic progress difficult.

Regressive pyramids are often more like pillars. They demonstrate low and declining fertility and mortality. This pattern is common in EMDCs where high living standards, education, and social awareness are usually accompanied by good food and health care.

Composite or **intermediate** pyramids are particularly common, and will show elements of two or more of the other model shapes. They display changes in fertility and mortality trends as countries pass through stages of development.

Age-sex pyramids reflect the social and economic character of individual countries - their state of advancement, the nature of their society, and even their future prospects. They also show aspects of the country's demographic history - wars, natural disasters and so on. In practical terms children and aged may be considered unproductive and so dependent on the wealth-producing adults. Clearly, the smaller the adult group relative to the other two, the more difficult it is for a country to be economically viable. However, one must note the complications which follow. In ELDCs it is more common for children to work at a young age. In EMDCs the reverse is true. Not only are children under 16 non-productive, but so are many young adults, because tertiary education is more widespread. At the other end of the spectrum, the aged group is becoming less productive in EMDCs as retirement ages fall. In ELDCs, however, this group may be more productive, necessitated by a usual absence of pensions. Migrations will be indicated by slight bulges (net inflow) or indentations (net outflow). Generally young adults are the most mobile. Finally, changes in the rate of population growth inevitably lead to changes in demographic structure. A falling growth rate results in an ageing population while a rising growth rate results in a youthful one.

Age structure may be considered collectively with the percentages of children, adults, and aged represented on a triangular graph - children plotted along one side, adults along a second, and aged along the third. When these are projected inwards they meet at one point. Many countries can be plotted on one graph allowing easy comparison. Likewise, the changing age structure of a country through time could be plotted.

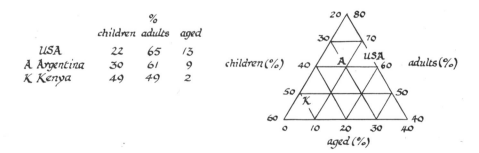

	%		
	children	adults	aged
USA	22	65	13
A Argentina	30	61	9
K Kenya	49	49	2

Sex structure likewise proves revealing. In broad terms, taking the whole population of an area, the two sexes should be equally balanced in numbers. However, overall ratios can hide differences within individual age groups. It is a proven fact, for example, that male births exceed female ones. The cause could be biological or social – relating to the widespread completion of families on the birth of a male child. It would be advisable to assume elements of truth in both unless specific evidence to the contrary is available. To balance this, women tend to outnumber men in the aged group because male mortality is higher.

In ELDCs male mortality exceeds female mortality particularly acutely. In Guinea, for example, the excess of male births is cancelled out within the first few years of life. Natural mortality is biased against males in all age groups for a combination of biological, environmental, and socio-economic reasons. Men are the 'weaker sex' in the sense that they tend to contract a greater number of illnesses. Likewise, war adds to male mortality, even when fighting more directly involves civilian populations, because men make up the bulk of the armed forces. World-wide, women work, on average, harder and longer than men, yet only occasionally does female mortality exceed that of males – such as where women have a lowly status and undertake heavy work, as in some parts of South-East Asia and Africa.

Finally, migratory patterns affect sex structure in that men tend to be more mobile. For example, male-dominated net migration is found in **pioneer regions** such as Alaska or the northern territories of Australia. Conversely, female domination is found where net emigration has occurred, as for example in the Irish Republic and the Caribbean. At a local scale, urbanisation in ELDCs will frequently illustrate massive male-weighting in resulting urban sex structure.

Occupational structure

Age-sex structure is a physical, innate characteristic just as race (ethnic composition) is. Occupational structure in contrast, is a social or acquired characteristic like language or religion.

Only part of a population is economically active, and normally quoted as a proportion of the total known as the **activity ratio**. As a general rule this will be higher in EMDCs where there are fewer children and more working women. Lower activity ratios in ELDCs reflect high juvenility and

the traditional non-employment of women, particularly marked in
Moslem societies

Undoubtedly there is a close relationship between occupational structure
and levels of development. Primary occupations, for example, dominate in
ELDCs. Tertiary and quaternary occupations, by contrast, dominate in
highly developed societies (see *Industry*).

Population problems

The study of world population growth quickly reveals that the so-called
population explosion is not a uniform global feature but effectively
concentrated in ELDCs. This feature is a particularly dramatic expression
of our divided world as analysed in the influential *Brandt Report* (1980)
which coined the labels **North** and **South** in order to overcome the
stereotyping associated with phrases such as 'Third World'.

North	**South**
EMDCs	ELDCs
'us'	'them'
'haves'	'have nots'
'affluence'	'want'
c. 20 per cent of humanity	c. 80 per cent of humanity
c. 80 per cent of wealth	c. 20 per cent of wealth

But these stark realities do not necessarily mean that population problems
are restricted to ELDCs. EMDCs also have population problems, most
often associated with so-called **greying (ageing)** age-sex structures.
Dependency ratios are, consequently, increasing with pension provision
funded by a relatively shrinking labour force. This is a typically pressing
economic issue, as is the funding and provision of appropriate medical and
social services. These, by definition, are without limit in their scope, and
present both moral and economic challenges.

REV. T.R. MALTHUS (1798) AND E. BOSERUP (1965)

Probably the first person to seriously consider the problems related to population growth was the Reverend Thomas R. Malthus in his famous 1798 *Essay on the Principle of Population*. One must be familiar with this work because his gloomy ideas have recently been resurrected by a **neo-Malthusian** school of thought.

Malthus examined the relationship between population growth and food supply, suggesting that whilst the former would rise geometrically (doubling every 25 years or so unless checked), the latter was likely, at best, to only increase arithmetically.

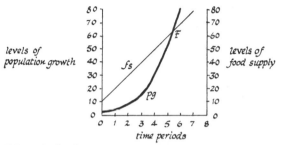

F beyond this point a time of famine   fs food supply (arithmetic growth)   pg population growth (geometric or exponential growth)

He therefore considered that the population increase should be kept down at a sustainable level by the operation of various checks - '*the positive checks of misery and vice*' in death by famine, war, and pestilence. He envisaged the chief preventative check to be '*moral restraint*', involving the deliberate decision by men to refrain from '*pursuing the dictate of nature in an early attachment to one woman.*' Malthus strongly opposed birth control within marriage and so anticipated smaller and probably fewer families solely by the delay. The **Malthusian League**, by contrast, strongly argued the case for birth control.

Malthus effectively envisaged a definite ceiling to food production in any area, so leading to less food per head until those with insufficient die, either directly from starvation, or from disease due to reduced resistance. This increased mortality, with related fall in fertility, acts as a **Malthusian check**.

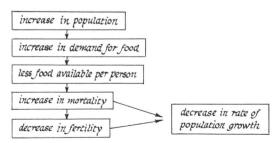

Looking further ahead, if the decrease in the rate of population growth led to a decline in total numbers to a level where food supplies were again adequate, the checks on growth would cease. Population would consequently increase again, stimulating a cyclical oscillation above and below the food supply ceiling as depicted by Haggett (1975).

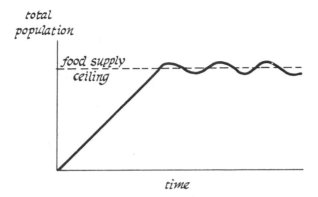

This pattern has been observed in animal populations and limited evidence suggests its applicability to primitive human societies.

NB: Malthus' gloomy predictions were not realised because some of his assumptions proved invalid. The Industrial and Agricultural Revolutions kept pace with demand whilst emigration to the New World acted as a 'safety valve'.

Boserup criticised the Malthusian approach because it was *'based upon the belief that the supply of food for the human race is inherently inelastic and this lack of elasticity is the main factor governing the rate of population growth.'* Population growth therefore depends on preceding changes in agricultural productivity. However, she suggested on the basis of research in ELDCs such as Bangladesh that population growth actually stimulates agricultural innovation - prompting the now famous quotation that *'necessity is the mother of invention.'*

So why have Malthus' ideas been resurrected?

There is evidence to suggest that the errors in his theory were primarily those of time scale and that the overpopulation and misery seen in so many ELDCs support his basic argument. However, the most pessimistic forecast of deaths due to AIDS, suggested by some as the ultimate Malthusian check, do not amount to a fraction of the c. 90 million extra mouths to feed each year as global population burgeons.

Indeed, the world is facing some harsh ecological facts given humankind's excessive demands upon the environment. There is an increasing imbalance between world population and material resources demonstrated by hunger and poverty in ELDCs and pollution, social service deterioration, and spoliation of amenities and natural environments in EMDCs.

Clearly political difficulties can follow in that regions or nations with enormous social and economic problems will undoubtedly be politically unstable. Examples of proliferating political parties, demonstrations, riots, and high crime rates are already apparent in Latin America and Africa.

Just as experiments with rats show their tendency towards self-destruction when their own numbers grow to endanger survival of the species, so it is likely that we are capable of likewise. Overcrowding in urban areas, for example, is associated with increasing vandalism and crime. Nationally it may be reflected in wars, as countries compete for limited resources. Understanding of resource provision and potential is thus critical.

THE CLUB OF ROME (1972)

Perhaps the best known attempt to assess the global situation regarding population growth and resource utilisation was by the interdisciplinary Club of Rome team, whose conclusions have been published by D.H. Meadows *et al* in *The Limits to Growth* (1972). They include the following:

1. '*If the present growth trends in world population, industrialisation, pollution, food production, and resource depletion continue unchanged, the limits to growth on this planet will be reached sometime within the next hundred years. The most probable result will be a rather sudden and uncontrollable decline in both population and industrial capacity.*'

2. '*It is possible to alter the growth trends and to establish a condition of ecological and economic stability that is sustainable far into the future. The state of global equilibrium could be designed so that the basic material needs of each person on earth were satisfied and each person has an equal opportunity to realise his individual human potential.*'

The first is arguably too alarmist in that the report failed to allow for appreciable changes in technology regarding conservation, recycling, resource substitution, and means of exploitation. The second hints that basic material needs are not currently met, yet could be, given the political will.

In conclusion, the environment does seem to have an ultimate **carrying capacity** - a ceiling beyond which extra numbers cannot be adequately fed, housed, or employed. The implications of this have been incorporated in the **exponential growth model** which suggests ways in which population growth could adjust to circumstances.

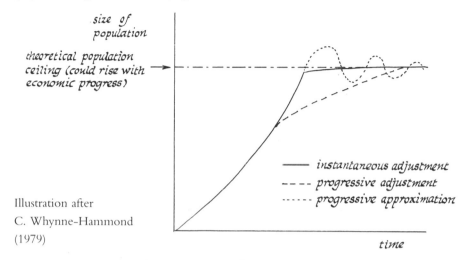

size of population

theoretical population ceiling (could rise with economic progress)

——— instantaneous adjustment
– – – – progressive adjustment
······· progressive approximation

Illustration after
C. Whynne-Hammond
(1979)

time

Instantaneous adjustment, where population growth suddenly becomes zero, is highly unlikely to occur in real life.

Progressive adjustment, where growth rate slows down before the ceiling is reached, is slightly more likely although it assumes that there is great control over the birth rate.

Progressive approximation, as in Haggett's 1975 graph, is the most likely, with periodic checks by famines, diseases and wars. This **J curve** is clearly neo-Malthusian.

Optimum population, overpopulation, and underpopulation

These refer to the relationship between population, the level of technology, and available natural resources. The **population** total or density represents a country's labour supply as well as its consumers. If labour productivity (output per person) is high, then consumption can also be at a high level and there will be comfortable living standards. **Level of technology** refers to the knowledge and capital equipment, such as infrastructure, possessed by the population. With a high level of technology a population will be more productive, so living standards can be higher.

Finally, **natural resources** do not determine standards of living. These depend upon the success with which a population applies its technology to use these resources.

There are not simple, logical correlations between population density and standards of living. Contrast Belgium with Bangladesh or Norway with Nepal, for example. Nor can simple relationships between resource endowment and living standards be made. Again, contrast Libya with Japan. Hence, whatever relationships exist, they are not straightforward.

Many **national population policies** are implicitly based upon the concept of an optimum (ideal) population. **Optimum population** is *'the size of population which allows the highest standard of living under present economic and technological conditions.'* Quite what criteria one uses to achieve a universally acceptable figure, is rarely expressed with any confidence.

Overpopulation, however, is usually clearly apparent. Defined as *'an excess of population over utilised or developed resources'* it may result from population growth or resource decline. It is characterised by low incomes, poor living standards, high unemployment, marked out-migration, and high population density.

Underpopulation, by contrast, invites far more debate. Compare Brazil with Australia, for example. Do both, one, or neither, represent *'a population which is too small to fully utilise its resources, or a situation in which resources could support a larger population without any reduction in living standards'*? Generally it is easier to define an overpopulated country than an underpopulated one. Both would have low living standards, but the former is more likely to have unemployment and underemployment, and a considerable amount of outward migration. Inward migration is more likely to occur with the latter.

Addressing population problems

A detailed appreciation of how nations tackle population problems will invariably cover numerous interrelated issues with economic, social, cultural, religious, and, not least, political aspects. In ELDCs **family planning programmes** will undoubtedly play a central role. However, **resource utilisation**, especially agricultural improvement schemes, may well figure prominently too. For example, agriculture globally could be extended beyond the c. 11 per cent of land surface presently cultivated by land reclamation, deforestation, irrigation, and marsh drainage. Initiatives relating to the so-called **Green Revolution** (see *Agriculture*) in selective breeding and the adoption of high-yielding, disease-resistant seeds, agricultural chemicals, hormones, and mechanisation could, with land reform, improve farming efficiency. Waste of food in transit and storage could also be reduced and systematic, controlled fishing of the world's seas and oceans provide far more protein than the lamentable c. 2 per cent managed by today's haphazard, destructive methods. But all too often this is potential rather than realised. A **food crisis** is clearly acute in ELDCs and all the more guilt-provoking when contrasted with the surplus production in EMDCs where advanced technology, plentiful investment capital, extensive land resources, and generally equable climates ensure that they can demand and obtain far more than their fair share.

11

MIGRATION

Migrations involve interactions between people and displacements of population - characteristics inviting a host of phrases and terms:

In- or **out-migration**, for example, refers to movements across internal boundaries whereas **immigration** or **emigration** cross international borders. Migrations are embarked upon from an **area of origin** and end in an **area of destination**. Migrants sharing common origins or destinations form a **migration stream** or **current**. The **counter-stream** refers to the lower volume reverse. When studying the effects of migration in a single area it is usual to refer to the total movement of population - immigration plus emigration equalling **gross migration**. The difference between the two, so relevant to population structure, is the **balance of migration (net migration)** - a positive or negative amount.

Types of migration

Unconscious drift describes the earliest wanderings of humankind, many originating from Mesopotamia - the so-called *'cradle of civilisation'*.

Compulsory movements, alas, predominate in contemporary migration streams - caused both now and in the past by economic need, and religious, ethnic, and political persecution. For example, between the fifteenth century start and nineteenth century conclusion of the notorious

slave trade, 10 million Africans were transported to the Americas. The events leading up to, during, and immediately following World War II, displaced 60 million in total. Migrations originally stimulated by Nazi persecution during the 1930s were multiplied by those fleeing fighting until 1945. Ethnic resettlements and labour movements following the peace were also significant. Contemporary **refugee movements** from various famines, civil wars, and disputed territories likewise number millions. Frequently, the immediate scale of these migrations defy comprehension. For example, over 1 million refugees crossed into neighbouring countries in order to flee **genocide** in Rwanda in 1994 – the largest recorded movement in recent history.

Voluntary movements may well, happily, involve greater numbers, but not as major streams. They allow us to consider the diverse geographical factors which determine human choice and behaviour. There must, for example, be very real physical, economic, political, and social reasons, together with knowledge, ambition, and energy stimulating the desire to move. Although nowadays movement is far more frequent and widespread due to improved communications and greater affluence, strong motives are still required.

Internal migrations, by involving relatively short distances, do not require the fundamental upheaval involved in, say, emigration. There should, for example, be few problems of adjustment to language, social customs, ideology, and institutions. **Urbanisation** and **counter-urbanisation** account for most movements. Indeed, as a general rule people become more mobile with the increasing affluence associated with economic progress. For example, around 20 per cent of the USA's population are thought to change address every year! Age is particularly relevant – the young moving for tertiary education, marriage, and careers, the middle aged to better housing, the aged to their 'retirement retreat'.

International migrations generally involve far greater distances and upheaval. Migrants may have to face a totally different physical and social environment – new climates, cultures, institutions, political systems, and even languages. Consequently motives must be stronger given that adjustments are likely to be problematic, with resulting integration, slower. Since the '*Great Age of Discovery*' 500 years ago, important international migrations have involved both **push** and **pull** factors discussed later, such as overpopulation, poverty, and hunger in the former, and higher wages, improved career opportunities, and better social facilities in the latter.

Indeed, such motives are still relevant today although the era of great world migrations is, arguably, over. More and more countries like Australia, the USA, and Canada now find that they can no longer absorb large numbers of aliens and so operate very selective immigration policies.

CIRCULATION

Periodic and **seasonal migrations** take place everywhere and may range in duration from days to several months. They are notably advantageous to all concerned - workers, for example, benefiting in monetary terms and economies remaining stable and healthy. Indeed, throughout Europe **guest workers** are crucial. In agriculture seasonal migration may be associated with the extra labour required at harvest time - such as with fruit gathering and potato picking in Europe. Likewise **transhumance** - the removal of livestock and farmers to high mountain pastures during the summer as winter fodder crops grow in the valleys, is a familiar Himalayan, and to a lesser extent Alpine, routine.

Daily migration may be associated with shopping and recreation, but is most relevant in **commuting** to work. This twice-daily routine has evolved over 150 years involving railways, trolley buses, trams, and, most significantly, the car – enabling pleasant rural and suburban homes to be enjoyed away from urban commercial and industrial workplaces.

Reasons for migration

Push factors encourage emigration by repelling migrants - **pull** factors encourage by attraction. Generally speaking, the stronger and more widespread these factors are, the greater distances migrants will be prepared to travel.

Each prospective migrant is vulnerable to perceptions distorted by anticipation - after all, '*the grass is always greener...*', a misconception often reinforced by **rose-tinted feedback** from earlier migrants, unwilling to admit the uncomfortable realities of their situation. However, 'weighing up' the advantages and disadvantages of present circumstances, as summarised below, is an essential prerequisite to the migration decision:

1. **Physical conditions** such as harsh climates, difficult relief and soils, natural hazards including volcanic eruptions, earthquakes, floods, droughts, and tropical revolving storms **push**, whilst agreeable climates, relief, soils, and so on, **pull**.

2. **Economic factors** such as unemployment, low wages, poverty, hunger, and malnutrition **push**, whereas the promise of higher living standards, valuable mineral deposits, and so on, **pull**.

3. **Social factors** such as changing family size and status due to marriage or children **push**, just like the limited entertainment opportunities and social activities so often cited in rural depopulation studies. Conversely, the urban 'bright lights' and expectation of independence, **pull**.

4. **Political factors** are rarely associated with voluntary migrations. However, past so-called Cold War defections from the now politically democratising eastern Europe were illustrative of this.

Theories of migration

Any migration decision is unique to individuals, yet people frequently act in groups and share in group decisions. When this occurs patterns may be identified, and theories and models formulated in an attempt at explanation.

THE LAWS OF MIGRATION
(E.G. Ravenstein, late 1880s)

1. Most migrants travel short distances and numbers decrease as distance increases – so illustrating **distance decay**.

2. Migration occurs in stages (**waves**) so that one short movement from one area leaves a vacuum to be filled by another short migration from beyond. In this way population progresses in waves towards an eventual goal.

3. Migration is a two-way process. Each movement has a compensatory movement in the opposite direction. Net migration represents the difference.

4. Where migration occurs over a long distance it tends to terminate in an urban area. The longer the journey the greater the likelihood of ending up in a major centre of industry or commerce.

5. Urban dwellers are less migratory than rural dwellers.

6. Women are more migratory than men over short distances, but men are more likely to move further.

These laws, based on Britain, have necessarily been abridged and modified in order to update and widen their still valid applicability. For example, increasing urbanisation in ELDCs, the predominance of young adults, the **stepped movement** rather than a sudden relocation from, say, village to major city, all supplement the original observations.

THE INVERSE DISTANCE LAW
(G.K. Zipf, 1930s)

This is a more sophisticated version of Ravenstein's first law and states that *'the volume of migration is inversely proportional to the distance travelled by the migrants.'*

$$N_{ij} \propto \frac{1}{D_{ij}}$$ where N_{ij} = the number of migrants from town i to town j

D_{ij} = the distance between the two towns

Clearly distance has a frictional effect on the volume of migrants. However, this does not recognise the pull exerted by the population of each settlement. In order to recognise this, various gravity models have been introduced.

THE GRAVITY MODEL
(based on Newton's law of universal gravitation)

This model assumes that a town's attraction is proportional to its size. Therefore, a town twice the size of another will have twice the pull and attract twice as many migrants.

$$N_{ij} = k \frac{P_i P_j}{D_{ij}^2}$$ where N_{ij} = the number of migrants from town i to town j

k = a constant to be stated

P_i = the population of town i

P_j = the population town j

D_{ij} = the distance between the two towns

THE THEORY OF INTERVENING OPPORTUNITY
(S.A. Stouffer, 1940)

This does not look at settlement size or the separating distance, but considers the perceived opportunities between them. It states that *'the number of persons going a given distance is directly proportional to the number of opportunities at that distance and inversely proportional to the number of intervening opportunities.*' Opportunities may be defined as pull factors such as employment prospects, vacant houses, and social facilities.

$$N_{ij} \; \alpha \; \frac{O_j}{O_{ij}}$$ where N_{ij} = the number of migrants between
                                                        town i and town j
                                              O_j = the number of opportunities
                                                        at town j
                                              O_{ij} = the number of opportunities
                                                        between the two towns

Migration theories and models appraised

All these models have a basis in reality. For example, data from England and Wales supports Zipf's theory by illustrating more than half of annual movements to be within local authority areas, with many of the remainder between adjoining regions and relatively few between distant places. Also, the gravity model is validated by the largest urban centres receiving a higher proportion of migrants than smaller cities which in turn receive a higher proportion than smaller towns. Therefore, it is a useful, if simplified, view of the reality of population movement. But it must be remembered that it only suggests that there exists a **potential** for movement to take place and not that it actually will. It takes the distance between centres as straight-line distance and does not take account of the variety or quality of routes between centres. It might, however, be more relevant to consider the time taken or the cost of travel. Nor does it take account of the intervening opportunities that exist between centres. Shoppers, for instance, may make purchases at local shops *en route* to major shopping centres, or a migrant seeking work might consider taking a job that he or she comes across *en route* to a major city. Hence Stouffer, therefore, adding the idea of 'intervening opportunity', where the volume of migration was directly proportional to the number of intervening opportunities, such as the number of job vacancies between the area of origin and the area of destination. In the gravity model, straight-line distance is measured between points, but one is essentially talking about areas of origin and areas of destination within which there will be variations in population density. Also, the gravity model takes no account of the factors pushing migrants out of their environments or attracting them to their destinations. Finally, population has been used as a substitute for mass, but more important substitutes might include average *per capita* income, the number of job vacancies, or the number of shops.

In conclusion, there has been a reaction, in recent years, against such abstract 'number crunching' approaches to migration by formulating more **behavioural studies**. These seek to understand the decision-making processes – not least why people decide to migrate and how they perceive alternative destinations.

Consequences of migration

Clearly migration will have significant effects on population distribution, composition, and growth. Areas of net emigration, for example, become depleted of young adults, especially men, and so reduced birth rates may result. Conversely, the opposite may occur in areas of net immigration. However, the consequences of migration are far wider ranging than just demographic structure. Movements of culture, technology, and ways of life are involved. Whole new landscapes may be created by design or accident. For example, potatoes from the Americas are now common in Europe, yet horses from Spain are more associated with the Americas. The rabbit, so common in Britain and Australia, is a native of neither! Likewise, architectural styles will represent similar translocations across the globe.

Economic results of migration are more difficult to identify and evaluate, but undoubtedly relate to capital flows which stimulate economic expansion in areas of immigration, whilst causing contraction in areas of net emigration. The centralisation of financial power in limited cities such as New York, Tokyo, Zurich, and London, arguably, illustrates this phenomenon.

However, many regard migration consequences as synonymous with problems and, alas, many are identifiable at all scales from, for example, local commuting, through regional urbanisation (see *Urban Settlement*) to refugee movements across international borders. Certainly the latter stimulates powerful emotions and reactions. The Rwandan illustration referred to earlier tested experienced aid workers and the media – not least newspaper readers and television viewers – to new extremes. Such was the scale of death from exhaustion, dehydration, and cholera in, for example, the Zairean border refugee camp at Goma that compassion fatigue was re-evaluated as '*a psychological defence mechanism*' due to '*information overload*', rather than callousness or insensitivity to the plight of thousands dying every day.

But it would be wholly misleading to regard the consequences of all migrations as problematic. Although the intermingling of diverse migrants has led to war, such as the late nineteenth century Boer War in South Africa, it can also engender greater understanding and tolerance between peoples. Indeed, one could argue that the universality of the English language, let alone our standards and institutions, have been a force for good. Not surprisingly, the social consequences of most migrations have both positive and negative aspects, as illustrated by plural (multi-cultural) societies.

Plural or multi-cultural societies

Despite being descended from the same source - probably Mesopotamia over 1 million years ago - our great variety of skin colours, hair forms, blood types, and so on, groups us into **races**. The crudest classification (Caucasian, Mongoloid, and Negroid) covers a great diversity believed to have originated from many generations of interbreeding within specific environments during humankind's long periods of isolation during successive Ice Ages. Lower sea levels allowed movements - hence Caucasians in Europe, the Middle East, and India, Mongoloids throughout the rest of Asia and the Americas, and Negroids, more commonly called negroes, in central and southern Africa.

However, these races are no longer solely associated with these locations. Since the '*Great Age of Discovery*', humankind has become more mobile - a mobility stimulated by various social, economic, and political factors and expressed in numerous large-scale migrations across the world's oceans. Such has been the development of plural societies in many countries that the distinction between races is now very blurred due to interbreeding. Indeed, race is arguably less meaningful than **ethnicity**, as defined by wider criteria including skin colour, nationality, language, religion, and culture.

In some instances, plural societies owe their origins to former European colonisation. Imperialism and foreign intervention have led to many former colonies containing four distinct ethnic groups - the indigenous population, the descendants of the European colonists, a non-indigenous, so-called 'coloured', population brought in as slaves, and people of mixed race resulting from interbreeding between the other three groups. North and South America, along with South Africa, offer examples of plural societies of this kind. The USA, for example, has relatively small numbers of indigenous Indians (Mongoloid), a great number of European

(Caucasian) stock, a fairly high percentage of African-Americans (Negroid), large numbers of Hispanics (especially illegal immigrants from Latin America), and a few racially mixed peoples like the Creoles of New Orleans.

PLURAL USA - A 'MELTING POT'?

The 'melting pot' is a well-worn but apt phrase, which describes the way in which a distinctive American society has evolved in a relatively short period of time from the merging of peoples from four continents. It is notable that the creators of the US set out, deliberately, to achieve a society in which political, social, and religious neutrality would produce a new and distinctive way of life - a new society free from poverty, class difference, and religious persecution. Most immigrants have readily adapted themselves to these ideals - anxious to prove themselves good American citizens. Indeed, hundreds of thousands did so to telling effect in the Union armies in 1861-65. But it is of great note, however, that at the same time, whenever a group of fellow-countrymen congregated, some remnants of their homeland culture would be kept alive through a church, home language, newspaper, or whatever. Generally speaking, the larger the alien community, the slower the assimilation and the stronger the nostalgic nationalism. Not so for the immigrants' children, however, enjoying wide social contacts with no language barriers to overcome. For them their parents' accent and cultural relics were largely curiosities. Schools deliberately overcame the alien influences within the home, producing a new generation of Americans pledging allegiance to the flag every morning.

Therefore, the obvious question is, of course - is it all happy and harmonious? Are the constitutional ideals realised? A closer look at the ingredients in the 'melting pot' will help to answer this.

The Vikings were the first Europeans to reach America, but Columbus (and Cabot) tend to get the credit for 'discovering' it. However, it was not until the seventeenth century that Europeans made settlements there. Most of these earliest European colonists were from Britain and Ireland and to a lesser extent Holland, Germany, and Sweden. What happened when these colonists took over the land of the indigenous Indians is told in the shameful stories of history. The consequences were expressed by their confinement to reservations - usually in unproductive arid and semi-arid western mountain areas. The seventeenth and eighteenth centuries also saw the slave trade whereby African Negroids were brought to work in southern plantations - tobacco in Virginia and Carolina, cotton in Alabama and the valley of the

Mississippi. Later immigrants have come from other countries including Italy, Greece, Russia (pre-1917 Revolution), China, Japan, and more recently from Indo-China (as a result of the Vietnam war), Mexico (1980s), Haiti, and Cuba (1990s). It is notable, however, that whilst Caucasian and Mongoloid races have, normally, few difficulties assimilating to the American way of life, Negroid African-Americans remain segregated by skin colour. This is worthy of closer examination.

The nineteenth century abolition of slavery meant that African-Americans in the southern states became share-croppers. They were free, but tied economically to the same tobacco and cotton lands as before and, of course, just as poor. This stimulated one of the most notable population movements of twentieth century North America - namely the migration of several million African-Americans from the farms of the south to northern urban areas, primarily the cities of North-East USA and the Great Lakes area. As a general rule they moved in search of work and to try to escape from discrimination. The industrial opportunities created by two World Wars gave even more impetus to the movement, such that Washington, the nation's capital, now has more than half of its population African-American. The result of this movement has been to involve the northern states in the reality of race relations that previously dominated the South - namely, can African-Americans compete for jobs on equal terms and should they be free to live in any street where they can afford a home? A legal framework has been created, with the ideal of leading to equality of status, in the **Civil Rights Act** - but in reality real equality is a long way off, despite the law, because of social pressure. African-Americans have tended to congregate in **ghettos** of old, cramped buildings, often near the city centre, which by pressure of numbers quickly deteriorate into derelict areas of blight. It is a tragic paradox that, while the equality of African-Americans has been legally reaffirmed on many occasions in recent years, concentration in ghettos has actually increased. Indeed, the ghetto is regarded by many as a refuge from racial discrimination.

The nature of the ghetto is well documented - supposed 'solutions' arguably less so. Rebuilding has been tried in New York and Chicago - derelict tenements have been torn down and replaced by new blocks of apartments. However, the social quality of this so-called solution is doubtful to say the least in that the basic problem still remains. Most African-Americans are still together - still segregated. Many argue that the only real solution is to break up the ghettos by enabling their inhabitants to disperse throughout the whole population, but this is just what poverty and social pressures prevent. Ghettos remain, therefore, as do residential suburbs around most cities with disproportionately few African-American residents.

12

AGRICULTURE

Agriculture is, arguably, the most important of all economic activities. By purposeful tending of crops and animals it produces food, industrial raw materials, such as fibres, and could, by further development of crop-based alcohols and oils, be an important energy source for the future. Agriculture provides the direct livelihood for at least half the world's people and its products form a major component of world trade.

In agriculture, humankind exploits ecosystems – in some cases to such an extent that the self-regulatory factors of the natural environment have had to be replaced by artificial regulation, using agricultural chemicals such as fertilisers and herbicides (weedkillers).

Agriculture tends to be the dominant user of rural land (compared to forestry, mining and quarrying, reservoirs, transport, settlement, recreation, and so on). The location and patterns of agricultural land-use at all scales, together with their changes, results from an interplay of physical, economic, cultural, and behavioural factors. The geographer's role is understanding and explaining these relationships, in order that this knowledge may be applied to useful purpose, such as in rural land management.

Although we have progressed from geographical determinism to possibilism, it must be accepted that environmental factors assert a major influence in determining the type of farming practiced. Indeed, a farmer's choice of land will be governed by the physical environment. However, technical and scientific developments can extend the margins of cultivation (limits of production) and environmental change can produce expansions and contractions in areas of, for example, arable (crop) farming.

Of the environmental factors affecting agricultural productivity, climate, soils, and relief are the most important.

ENVIRONMENTAL FACTORS INFLUENCING AGRICULTURE

1. **Temperature** dictates the growing season as determined by the number of frost-free days.

2. **Precipitation** determines the water supply in its widest sense.

3. **Wind** and **storm frequency** restricts cultivation of grain crops.

---

4. **Soil quality** is fundamental and determined by factors such as depth, texture, structure, mineral content, pH, aeration, capacity to retain water, and vulnerability to leaching.

5. **Relief variables** of altitude, gradient, and aspect interrelate with all of the above.

In temperate climates, a growing season exceeding three months (with mean annual temperatures above 6°C) is required with annual precipitation between 250 and 3000 mm and potential evapotranspiration of less than 375 mm.

Soils, altitude, and slopes also impose limits. Soils influence crop production by their supply or deficit of soil moisture, together with their type and stock of available nutrients. In Britain, for example, potatoes fail if acidity falls to less than pH 4. The upper limit for hay and potatoes is 300 m and slopes of more than 11 degrees become impractical for safe ploughing.

It is argued that there is an optimum or ideal location for each specific farming system according to climate, soils, slopes, and altitude. I I.H. McCarty and J. Lindberg's (1966) **optima and limits model** demonstrates the progressive deterioration in conditions with increasing distance from the most favourable (optimum) location – with inevitable operation of the law of diminishing returns.

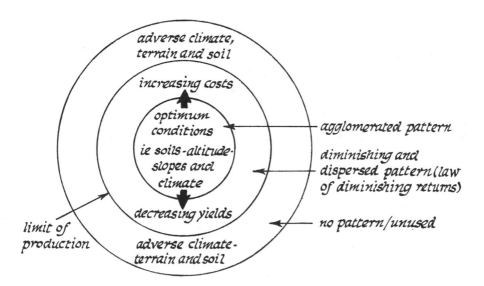

After D. Waugh's (1990) adaptation

The **law of diminishing returns** states that successive applications of labour and capital to a given area of land will, ultimately, other things remaining the same, yield a less than proportionate increase in produce. There is a limit, for instance, to the yield from a crop with successive applications of fertiliser.

The same law applies to successive extensions of agricultural land for a particular type of land-use. There comes a point when it is no longer worthwhile for a farmer to bring more land under production since he is incurring rising costs yet decreasing yields. This point is known as the **margin of cultivation.**

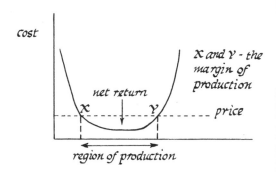

Such models, by simplifying reality, have a valuable role to play in aiding our understanding of such a wide and diverse subject as agriculture. Indeed, the first geographical model ever published was seeking to understand the location of agricultural production.

J.H. VON THÜNEN'S MODEL OF AGRICULTURAL LAND-USE
(*The Isolated State*, 1826)

This remains the most famous of all studies of agricultural land-use. It is based upon von Thünen's experience of farming in Mecklenburg on the Baltic coast of Germany. Indeed, most of the data he used to explain his theory was obtained from this practical experience - including detailed cost-accounting on his estate.

The book had two main aims - to demonstrate both **how** and **why** agricultural land-use changed with distance from a market.

In order to do this he presented a theory of land-use around a market which had as its fundamental principle the idea of locational rent.

Locational (economic) rent is the net return (**profit**) gained by a farmer from an area of land given over to a particular crop or farming system. This was determined by the production costs per hectare, market price per unit of commodity, transport costs, and distance from market.

Like all models in geography, von Thünen's made several basic assumptions in order to simplify reality. The world in which his land-use evolved was presented as an 'isolated state' in which no external trading took place and one central city was the only source of supply and demand. The area was a flat, featureless (isotropic) plain in which there was no variation in soil fertility, relief, or climate and on which movement was of equal ease in every direction. There was only one form of transport (horse and cart) and transport costs were directly proportional to distance – so illustrating the **friction of distance**. All farmers received the same price for a particular crop at any one time and were considered to be eminently rational **economic men** aiming to maximise their profits whilst having perfect ability and full knowledge of market demands.

These assumptions allowed the basic theory to be formulated. The assumptions could then be progressively relaxed in order to allow it to conform more to the complex real world situation.

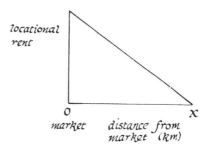

If we take a simple case, farmers are growing the same crop, receive the same price, have the same production costs and yield per hectare, and have equal accessibility to market. Locational rent (LR) is therefore maximised at 0 and falls to zero at x which represents the margin of cultivation.

Using the following formula and data, the locational rent for a particular crop at increasing distances from the market can be shown:

$$LR = Ym - Yc - Ytd$$
$$= Y (m - c - td)$$

where **LR** = locational rent per unit of land
       Y = yield per unit of land (hectare)
       m = market price per unit of commodity
       c = production cost per unit of commodity
       t = transport (freight) cost per unit of commodity
       d = distance from market

m, c, Y and **t** are constants – they do not vary. The dominant factor or independent variable is **d**.

The LR for this particular crop – yielding 1000 tonnes/km^2, fetching £100 per tonne at the market, costing £50 per tonne to produce with transport costs of £1 per tonne/km can, therefore, be illustrated:

Distance from market	c	Y	td	m	LR
0 km	50	1000	0	100	£50 000
10 km	50	1000	10	100	£40 000
20 km	50	1000	20	100	£30 000
30 km	50	1000	30	100	£20 000
40 km	50	1000	40	100	£10 000
50 km	50	1000	50	100	£0

Von Thünen considered distance and, consequently, transport costs as the dominant factors accounting for changes in land-use and more or less intensive farming systems. Again, these were studied with respect to the central, urban market. The location of the particular land-use or farming system would depend on its yield per hectare, the character of the product in terms of its bulk and perishability, production cost per hectare, and market price.

So far only one crop has been considered. Where a variety of products, crops, or systems of production are involved, the crop, or system yielding the highest LR (profit) will be found closest to the market, whilst others will be found at particular distances from the market where they maximise their LR by producing commodities for which the location relative to the market offers them the greatest advantage.

Von Thünen considered that the intensity of production for a particular crop would decline with distance from the market.

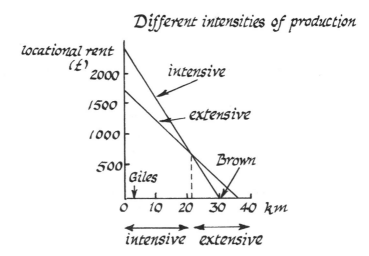

Different intensities of production

M.G. Bradford and W.A. Kent (1977) illustrated this effectively by considering the effect of distance from market on farmers Giles and Brown, given a market price of £55 per tonne and transport costs of £1 per tonne/km:

	Intensive farming		Extensive farming	
	Giles	Brown	Giles	Brown
Distance from market (km)	1	30	1	30
Production costs (£)	2000	2000	1000	1000
Yield (tonnes)	80	80	50	50
Total transport costs (£)	80	2400	50	1500
Total costs (£)	2080	4400	1050	2500
Total revenue (£)	4400	4400	2750	2750
Locational Rent (£)	2320	0	1700	250

The **comparative advantage** of farmer Giles is greater if he adopts intensive farming methods. The disadvantage of Brown is less if he farms extensively. Indeed, it is the only way he can make a profit.

Therefore, intensity of land-use (costs of production and corresponding yield per hectare) generally decline with increasing distance from the market because, as yields increased, transport costs per unit of product did likewise. Consequently, at a certain distance from the market, returns from higher yields do not outweigh transport costs. Using an intensive system LR falls to 0 at 30 km - yet using an extensive system, with resulting lower yields, at a distance of 30 km savings in production costs (man hours and so on) are more than the increased transport costs. Beyond 21 km, therefore, extensive farming is the most profitable system.

Von Thünen did, however, record exceptions to this general rule. For example, timber was found close to the city - even though production costs were low – because it was very high yielding per hectare, bulky and costly to transport, and commanded a low market price. Butter production was found at some distance from the city - even though production costs were high. This was because it was a product of high value and small bulk. It was therefore able to stand the cost of transport.

Von Thünen then considered how the type of land-use would vary with distance from the market. Clearly, in this example, potatoes produce a larger bulk per hectare and command a lower market price than wheat. Also the LR for potatoes decreases more rapidly with distance. Potatoes will, therefore, be grown closer to the market.

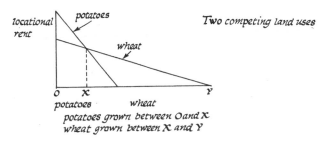

locational rent

potatoes

Two competing land uses

wheat

O   x   y

potatoes   wheat

potatoes grown between O and X
wheat grown between X and Y

Von Thünen finally translated the LR for various land-uses and systems of production into a model of land-use around one central city. Note that the right hand side of the diagram includes further modifications made including improved means of transport (a navigable river). The cheapened transport allowed zones to extend along the river, producing a linear form rather than the concentric circles of the original. A subsidiary town with its own trade zones was also introduced.

Von Thünen's model of land use around one central city

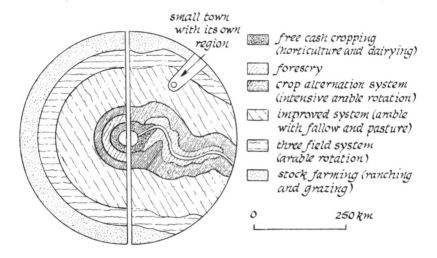

small town
with its own
region

- free cash cropping (horticulture and dairying)
- forestry
- crop alternation system (intensive arable rotation)
- improved system (arable with fallow and pasture)
- three field system (arable rotation)
- stock farming (ranching and grazing)

0   250 km

Von Thünen did allow some other aspects of the real world to be included and so alter his model. He later allowed differences in soil fertility and climate to affect production costs in his landscape. However, he did not allow variations in terrain which is very influential in reality. Significantly, he later accepted differences in the behaviour of farmers, relaxing the idea of profit maximisation by considering the effects of differences in management. He also proposed that farmers would, in reality, differ in their views as to what was acceptable profit.

Land-use zonation in the real world can be illustrated with varying degrees of success at all scales – local, national, and international, albeit distorted by the complexities of reality. Von Thünen could not of course foresee changes in transport technology such as refrigeration, which some suggest merely enlarge the zones. Certainly transport costs vary with direction and mode and are not directly proportional to distance - but they are nowadays of decreasing relevance. Consequently, physical factors such as climate, soils, and relief have become more important, leading to increasing agricultural specialisation.

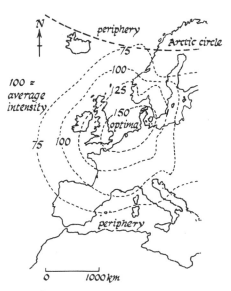

Intensity of agriculture in Europe – Von Thünen's model applied at an international scale

Appraising von Thünen

The model has been criticised as static - not taking into account the effects of urban expansion encroaching on the surrounding farming land. Indeed, this expansion causes land at the urban fringe to lose its relative value for agriculture as farmers, fearful for their future security, invest less effort in it and are increasingly tempted to sell to developers.

W.B. Morgan (1973) is particularly critical in *The Doctrine of the Rings*, stating that von Thünen's theory *'depends on assumptions which are in part incompatible with the basic argument.'* Morgan suggests, for example, that since manuring declined with distance from the city the inner ring had, therefore, higher fertility - so breaking the model's basic assumption of equal fertility throughout. He also questioned the ring of forestry, pointing out that timber in reality would have differing values for fuel compared to building and that high value wood for the latter was found further away and transported cheaply by water.

Certainly, in the real world, price changes and the absence of perfect competition make von Thünen's static model unworkable. Indeed, where ring patterns do exist they may, as Morgan suggests, result from innovations induced by the urban centre rather than increasing transport costs with distance. Also other land-use patterns can be identified, but von Thünen recognised this in stating that *'the farming systems would not (in reality) succeed each other in a regular succession, as in the isolated state, but would be jumbled up among each other.'*

Perhaps the main problem, however, is the assumption of economic man. It is unreasonable to say the least that a farmer could be an eminently rational profit maximiser with perfect knowledge about existing and future conditions - or, indeed, perfect ability to use the knowledge! It is highly unlikely that optimum decisions are ever made. Remember - the physical, social, and economic environment offers the farmer possibilities which he will see in the context of his background, society, and personal goals. These may be balancing income with sufficient leisure time, or, as is often the case, seeking prestige amongst peers. Indeed, recent research on decision-making and perception of the way individuals see their environments is of particular note here. A study of central Sweden, for example, compared actual to potential productivity. It demonstrated only 40-70 per cent of the potential being achieved because perfect knowledge was unattainable due to uncertainties in prices, costs, the weather, and so on - let alone varying access to information. Another study, in the Great Plains, USA, concentrated on how well farmers perceived environmental hazards. It concluded a farmer's efficiency in anticipating drought and hail was dependent on the intensity and frequency of the hazard. Greater awareness was found in areas where they occurred most frequently and was related to experience - the most experienced farmers being most aware.

Any consideration of von Thünen would have to address the often repeated criticism that it is anachronistic and bears no resemblance to contemporary economics. But remember - von Thünen himself went to considerable lengths to point out that his work was essentially a method of approach to the undoubtedly complex subject of agricultural location. He openly admitted that his findings had no claim to universality - but the methods by which they were obtained could be applied generally.

Perhaps the most neglected aspect of the model is that of its scale. M. Chisholm has suggested that the principle may be applied to the land-use on a single farm, or around a village, or even to patterns at national and continental scales. Indeed, it is notable that Bradford and Kent's contemporary examples are examined at three scales - farm, village, and continental.

My conclusion is, consequently, based heavily upon their recognition of von Thünen being useful in two main ways:

1. He focuses attention on economic factors, particularly transport costs and distance to market. (Remember, geographers in the past have always treated these as far less important than the physical environment when studying land-use.)

2. He introduces the concept of locational rent into geography – a concept useful when studying urban land use as well.

It would seem that the rigid assumptions of the model have led, indirectly, to an increasing research emphasis on farmers' decision-making processes, the quantity and quality of their information, and their willingness or otherwise to innovate. As a result a greater understanding of rural land-use is slowly but surely emerging. So perhaps we should view von Thünen's work, with all its limitations, as a useful framework for organising, for example, village and farm studies. Certainly, Bradford and Kent's argument is that most models emphasise **residuals** (unexplained parts) and these tend to be of greatest interest – so leading the student to greater understanding. This must be regarded with respect.

Illustrations adapted from Bradford and Kent (1977)

Economic and cultural factors affecting agricultural land-use and profitability

Labour inputs (and capital inputs outlined later) are also subject to the law of diminishing returns, particularly in ELDCs where overpopulation often results in the underemployment of labour and fallow periods being reduced. Inefficiency can be related to disease (especially malaria), imbalanced diets, juvenility, and land tenure systems and inheritance laws which have produced small farms and highly fragmented plots.

In commercial agriculture, labour costs can be a significant factor. Indeed, an important problem in EMDCs is that of matching the seasonal rhythms of agriculture to labour employment – particularly in **monoculture** systems whereby the same crop is cultivated each year on the same piece of land. In temperate regions there is a notable disparity of incomes between agriculture and industry, as well as a great variation in the length of working week. Traditionally low wages and increasing mechanisation have led to a drift of labour from the land, especially of young people.

Capital inputs may be difficult to secure because farmers find attraction of capital and credit facilities much more difficult than the industrialist because of the greater uncertainty in agriculture. In ELDCs, for example, constant debt is not unusual, especially in marginal areas where crop failures are common. Money-lenders, such as the *bania* of India, charging interest rates in excess of 50 per cent are not uncommon and rents claim up to half of the gross value of output. To buy machinery can be risky given the uncertainties of environmental and economic conditions - risks which can, however, be reduced through the adoption of cooperative schemes for equipment purchasing and product marketing.

Cultural forces play a major role in shaping both land-use and output. Subdivision of land as a result of inheritance laws is probably the most obvious expression of this. In much of Africa, for example, the land tenure system is one of female inheritance. Males in the community often have little interest or incentive, therefore, to improve the land. In savanna regions the prestige associated with pastoralism and the size of herd often leads to overgrazing and soil deterioration - hence exacerbation of desertification in West Africa. Another example of cultural forces is in South-East Asia, where the early diffusion of more advanced agricultural techniques saw neighbouring cultures taking different steps - either accepting or rejecting them. The early Malayan (Malaysian) rejection of terraced rice cultivation, adoption in Indonesia and the Philippines, and the non-acceptance in New Guinea and the rest of the Pacific, are often quoted examples. Although rather dated, Malaya's (Malaysia's) Federal Land Development Agency's (1965) observations on the progress of land colonisation schemes demontrate cultural forces admirably in finding that '...*pioneering efforts showed a success dependent on varying ethnic groups rather than advantages in physical conditions.*'

Government intervention not only influences, but can control in some cases the pattern of agricultural land-use and efficiency. Collective farms in the former USSR were the ultimate expression of this. In capitalist economies, however, the more usual forces are those of quotas, the policies of marketing boards, the use of subsidies, or voluntary land reform schemes. The latter redistribute land to the landless or small farmer by breaking up large estates, such as in Italy's Mezzogiorno, or consolidating small farms and parcels of land into viable economic units, such as in France following World War II.

Agricultural development and change

Before studying specific farming systems it would be prudent to reflect upon the ways in which agriculture changes through time and to identify any prevailing misconceptions and stereotypes. Subsidised overproduction associated with the **Common Agricultural Policy (CAP)** of the European Union (EU) is much quoted - but rarely with balancing references to contemporary initiatives, such as **set-aside**, which address the problem. Indeed, such is the assumption of enviable productivity in European **agribusiness** that it tends to disguise the fact that towards the periphery there still persist fragmented, inefficient, and undercapitalised farms. Today represents a much-improved situation from the 1950s when 20 per cent of western Europe's farmland was in need of plot consolidation. In spite of the difficulties in identifying optimum farm sizes (given that, say, cereals would demand more land than market gardening – let alone the environmental vagaries of different conditions) one can safely assume that the standard governmental response to small, inefficient farm units involves the stimulation and encouragement of cooperatives, provision of grants and low interest loans for land, property, and equipment improvements, the setting up of advisory schemes, and direct intervention through price support.

Studies of land reform in southern Italy (the Mezzogiorno) and Chile demonstrate interesting parallels which challenge many stereotypes about EMDCs contrasted to ELDCs. Their problems were primarily social and political – but both have a physical geography dimension.

LAND REFORM IN THE MEZZOGIORNO AND CHILE

The **Mezzogiorno** had a dual farming economy and society in the 1950s, with 40 per cent of its production agricultural. Large estates grazing goats and cultivating olives, citrus fruits, and vines were inefficient, undercapitalised, and often owned by absentee landlords. This contrasted with a large group of peasant share-croppers and landless labourers living in discontent, poverty, and ill health – often in remote hill villages served by a limited infrastructure of poor roads, telecommunications, water supplies, drainage, and sewerage systems. The physical environment was also problematic. Relief, for example, is rugged – with limited silty, often marshy (coastal) flat land. The Mediterranean climate is seasonal with high temperatures and limited rain coming in winter downpours following the summer drought. Thin limestone soils are vulnerable to wind erosion in summer and gully erosion in winter. Deforestation and especially overgrazing were endemic and marketing at best ramshackle and often blatantly corrupt. With 30 per cent of the total labour force in agriculture, land reform was essential.

Government capital on a massive scale was invested in a special 'Fund for the South'. The resulting CASSA schemes invested in infrastructure developments, land reform agencies, and appropriate industry. Soil conservation programmes, hillside terracing, afforestation, marsh drainage, reservoir construction, and irrigation schemes were embarked upon. Cooperatives were set up for marketing and bulk buying of machinery, seeds and agricultural chemicals. Processing factories, agricultural advisory centres, schools, hospitals, and so on completed the infrastructure improvements. However, central to the land reform schemes was the compulsory purchase of large inefficient estates and their subsequent subdivision in order to create small intensively cultivated family farms for previously landless peasants. The new owners could then be educated in farm management by residential advisers.

Appraisals of land reform in the Mezzogiorno range from 'very successful' to highly critical. Many have observed that the problems of the small farm, such as in France, have been created - and often exacerbated by lack of water and awkward relief. Certainly inexperience and a new isolated lifestyle, divorced from the traditional village community, led to many men abandoning the challenge and migrating north for industrial work - so leaving their families behind to tend the land.

Chile is known infamously for its long history of human rights abuses and very marked social contrasts. Physically it is also a country of extremes - with extraordinary climatic diversity dictated by its linear form following the western margins of South America. The hot, dry Atacama desert in the north contrasts markedly to the cold, windy, and wet southern zone. It is not surprising, therefore, that agricultural, industrial, and political power is concentrated in the central provinces of Santiago and Valparaiso which enjoy a temperate Mediterranean climate. Livestock, wheat, maize, potatoes, sugar-beet, fruit, and wines represent an important component of Chile's domestic and export trade. Production has traditionally been dominated by big estates (haciendas) worked, but not owned, by large numbers of peasants.

Tentative attempts at land reform to increase agricultural output, settle peasants in their own farms, and alter Chile's rigid social stratification began in 1928 with the setting up of an Agricultural Colonisation Bank. But it took the establishment of a Corporation for Agrarian Reform (CORA) in 1962 to actually acquire land and set up cooperatives on abandoned estates. However, real change came in the late 1960s with the democratic election of the Marxist Allende who, understandably, considered land reform for the people a priority despite the problems of small unirrigated farms quickly becoming apparent - just as in southern Italy. Reform, however, following the brutal military coup in 1973 which installed 'President' Pinochet as dictator, has been minimal, lending credence to speculation that American based trans-national companies (including IBM and Coca Cola), which owned many haciendas, were implicated with the CIA in Allende's overthrow.

A new constitution has recently returned democracy to Chile. Its economy continues to be dominated by primary products and it remains to be seen whether or not significant agricultural land reform regains momentum.

Further agricultural development and change centres round **land extension** and the Green Revolution (referred to in *Population*).

Deforestation, hill land improvement, marsh drainage, reclamation from the sea, and, most significantly, irrigation have been enormously influential in extending the margins of cultivation. The southern edges of the boreal forests of the CIS now support hybrid strains of wheat, barley, and rye on drained and fertilised podsol soils; the drained heaths and boglands of central Ireland and Scandinavia are now productive, and the polders reclaimed from the North Sea in the Netherlands are all much-quoted examples of successful land extension. However, agricultural development through **irrigation** is particularly dramatic. Political and strategic motives, for example, lie behind much of the 'greening' of Israel. 60 per cent is arid, yet the *kibbutzim* (collective villages) and *moshavim* (cooperative farms) harness potentially fertile loess soils entirely through irrigation. The Central Valley Project of California is another excellent example, given the agricultural productivity (and urban excesses) in this predominantly arid region. Yet all irrigation schemes face problems of evaporation and salinisation, which require careful management in order to avoid salt sterilisation of vast areas – and original water supply may not always be effortless. Multi-purpose river projects such as Egypt's Aswan High Dam scheme may have spectacular potential, but are not without problems – not least the displacement of existing population from the planned reservoir site. Desalinisation of sea water represents a very expensive option for 'oil rich' Arab states, yet direct use of saline water, as pioneered in Israel and Tunisia, may well offer hope for the future. Perhaps most potentially rewarding, however, are the vast, as yet untapped supplies of underground water globally. Only c. 20 per cent of India's aquifers, for example, are currently used.

Finally, the Green Revolution has undoubtedly generated much optimism since the 1960s. Modern technology applied to mechanisation, irrigation, soil conservation, crop storage, handling, and processing – and especially research into new chemical fertilisers, fungicides, and herbicides, in addition to plant hybridisation and selective breeding – have undoubtedly improved agricultural productivity. But the pioneering efforts of Doctor Norman Borlaug and other agronomists proved to be no universal panacea – they have not eliminated hunger as early forecasters prophesied. The high levels of capital investment by national governments necessary must also involve education of farmers traditionally sceptical of innovation. After all, it is they rather than the scientists who suffer if, for example, high-yield (seed) varieties (HYVs) fail through misapplication of the necessary chemicals. Similarly, mechanisation requires support services for fuel and repair which are not always widespread or, paradoxically, welcomed, given the job losses associated. Whilst examples abound of successful innovation, it has proved inescapable that the poorest farmers, which form the majority in ELDCs, have gained little. Indeed, the social and environmental costs have been marked, not least the widening gap between those who can afford such innovation and those who cannot.

Agriculture in ELDCs

In ELDCs agriculture is very important, often occupying over half of the labour force. Most, although not all, is **subsistence farming** in some form or another, and each farm community is largely self-sufficient. Indeed, there tends to be limited specialisation (excepting plantations and cash cropping share-croppers). Certainly, where trade is limited, each farmer would have to produce a wide variety of, predominantly food, products. Production depends to a large extent on human labour, assisted sometimes by animals. Capital for mechanisation and other innovation is scarce, so productivity per worker is usually low. However, in areas of intensive farming, productivity per hectare may be high.

Briefly, nomadic herding exists mostly in the arid belt extending from North-West Africa to central Asia. Primitive subsistence agriculture (including shifting cultivation and bush fallowing) is associated with the hot, wet equatorial regions of South America, central Africa and the East Indies. It is in these areas that Europeans have made comparatively little impact upon farming apart from the development of plantation agriculture and irrigation schemes such as the Aswan High Dam project. Finally, intensive subsistence agriculture is found in East and South-East Asia.

Nomadic herding is an **extensive** system of agriculture whereby there is little, if any, capital investment and relatively few people are supported by large areas of land. It is virtually confined to the Sahel region of Africa and South-West Asia (where the first animals were domesticated about 8000 years ago). However, reindeer herders do still exist in the tundra regions of Eurasia. (It is of note that when Europeans first visited the *New World* of the Americas and Australia they encountered no animal herders. It is reasonable to assume, therefore, that the idea of domesticating animals never spread beyond the *Old World*.)

In nomadic herding the herders move between grazing grounds according to the season of the year. Their animals provide milk, cheese, butter, meat, and hides as well as transport. Usually there is no permanent village and the whole community may migrate with its herds, carrying tents and other provisions. Another distinctive feature of the system is that there is no private ownership of land – the social group identifies itself with its leading members rather than with a particular area of land. Usually, there are no clear limits to the areas of migration, which will necessarily extend in poor years and always be influenced by seasonal climatic variations. It is of note that nomads are frequently an embarrassment to national governments because they have no settled homes. As a result, some attempts have been made to settle them in a process known as **sedentarisation**.

THE FULANI OF WEST AFRICA

The Fulani are found from southern Mauritania to northern Nigeria. In the latter, the tribes graze long-horn cattle across the savanna grasslands north of the River Benue. Their animals provide food (dairy products and meat), clothing and shelter (hides), and dung for fuel. Fulani migrations cover hundreds of kilometres and follow the seasonal rains - hence pasture. These come earliest in the north but last longest in the south. The men leave their wives and children near the River Benue in April and migrate northwards following fresh pasture to the Jos Plateau for the summer. Here they set up a semi-permanent settlement. At the end of the rainy season they return south to rejoin their families who have been tending subsistence and fodder crops of sorghum, millet, and maize. The range of migrations is restricted by tsetse fly infestations in the south and, ultimately, the Sahara Desert to the north.

The Fulani lifestyle is, inevitably, changing. Population growth, and persisting cultural traditions perpetuating large herd sizes as a measure of status, exacerbate overgrazing and desertification. As grazing becomes more scarce, so the potential numbers supportable is reduced. Also, now that dairy products are traded, clothes, radios, and so on can be bought. Sedentarisation, therefore, through both environmental pressures and commercialisation, may be inevitable - but only realistic if the governments of Nigeria and other Sahel countries encourage and invest in rural development programmes to include well drilling, agricultural and general education, family planning, and so on.

Primitive subsistence is also an **extensive** system of agriculture. It includes shifting cultivation in which the cultivated plots are abandoned every few years and sedentary bush fallowing in which the same plots are rotated.

Shifting cultivation was once common in Europe and also practiced by North American Indians. However, it is now effectively confined to tropical rainforests and, occasionally, wooded savanna grasslands. It has various names depending on the location – such as *milpa* in Latin America, *chitimene* in central Africa, and *ladang* in South-East Asia.

AMERINDIANS OF THE AMAZON BASIN, BRAZIL

Amerindians are rapidly declining, as commercial development of the Amazon rain forest threatens their traditional lifestyles. Tribes such as the Boro, still practising shifting cultivation (*milpa*), have been forced into particularly remote areas of the forest west of Manaus. Here, however, the world's simplest farming system still involves felling several small clearings with crude stone axes and machetes at the least rainy time of year around August. All but the largest trees are cut to shoulder height, and ground cover is also cleared. Branches and foliage are then left to dry before being burnt – a process which also kills weeds. The fertile ashes are subsequently spread amongst the living stumps prior to wooden digging sticks being used to plant manioc for cassava bread, yams, pumpkins, peppers, beans, and occasionally maize. Hand weeding is necessary because both crops and weeds grow rapidly in the hot, wet conditions. Cultivation can continue for around three to five years before the clearings lose their fertility and force the tribe to move on – abandoning their crude dwellings to natural regeneration of the forest. Providing they do not return for up to thirty years, the rainforest will recover. This **slash and burn** system may seem inefficient and wasteful of land – but it has proved sustainable over generations, limiting soil erosion to a minimum in what is proving a fragile environment if cleared for permanent agriculture, such as ranching, or for timber and mineral exploitation.

Bush fallowing represents an intensification of shifting cultivation in that settlements become permanent and only the cultivated plots are abandoned and recreated in a simple **land rotation**. Moderate, rather than low population densities can be supported – based, for example, on maize in the Andes of South America or sorghum in the plateaux of central Africa.

Intensive subsistence agriculture has for many centuries been a characteristic of the valleys, coastal lowlands, and hillsides of East and South-East Asia – the staple crop being rice because it is nutritious and stores well. Productivity per hectare can be very high allowing rural population densities up to 2000 per km^2 to be supported. However, productivity per worker is low, with consequent poor living standards. The Green Revolution's 'miracle rice' (IR8) has been introduced in many areas, including India, with varying degrees of success – usually dependent on the farmer's ability to afford the necessary fertilisers and irrigation control.

RICE CULTIVATION IN INDIA

The scale of Indian agriculture is difficult to comprehend. That in area, and especially workers, it dwarfs any European country is obvious – but twice the area and twenty times the workers of the EU in total, puts it into impressive perspective. However, unlike Europe, most Indian agriculture is not commercial, but subsistence farming, albeit intensive, producing high yields per hectare through many people working in the fields. The staple crop of rice is ideally suited to the region's monsoon climate. Rice requires at least three months with temperatures above 20°C, abundant water for the padi fields, and a dry season for the crop to ripen. India's climate is ideal – and contrasting environments provide fertile soil for often two crops per year. 'Wet' rice is cultivated in the thick, fertile alluvial silt of the Ganges flood plain. Elsewhere, 'dry' rice is grown, irrigated on steep hillside terraces, as in Assam, or on the fertile volcanic soils of the Deccan Plateau. Whatever the location, the process starts before the monsoon rains begin in May. Nursery seed beds are prepared and manured. Once the rains start the seeds are planted and carefully tended for forty days. Meanwhile the padi fields are levelled, manured, and the perimeter banks (*bunds*) checked carefully before flooding. Hand transplanting of individual seedlings, in July and August by women and children, is followed by rapid growth and constant weeding up to harvest in drier conditions towards the end of the year. Drying of the cut plants is followed by threshing, winnowing, and, after further drying out, storage. In January, providing irrigation water is available, a second crop may be grown for harvest in April. However, often other crops, such as beans, lentils, or wheat are cultivated.

Plantation agriculture is one of the chief ways in which EMDCs have influenced cultivation in ELDCs. Monoculture of tea, coffee, cocoa, cane sugar, bananas, rubber, palm oil, sisal, and so on, provide cash crops in bulk for export throughout the world – so accounting for frequent coastal locations. Highly **intensive** in both capital and labour, this **tropical commercial** agriculture is notable in that although now often owned and managed by EMDC-based trans-national companies, it is regarded by many as exploitative and an undesirable symbol of neo-colonialism. Indeed, colonial origins in the seventeenth and eighteenth centuries, not least its association with the slave trade, certainly cannot be ignored. Nor can other significant population migrations associated with this form of agriculture – such as **indentured** Indian labour to South Africa, Malaya (now Malaysia), and Ceylon (now Sri Lanka). However, there are many benefits associated with this form of agriculture – not least the employment opportunities, higher standard of living and housing, shops, schools, clinics, and utilities often provided for the workers and their families. Nowadays, subsistence farmers have frequently taken up the production of plantation crops as **share-croppers**.

SISAL PLANTATIONS IN TANZANIA

Sisal provides a fibre from its leaves and stems which is used in the manufacture of ropes, twines, sacks, matting, and strong papers. The details of its production are less important than the excellent example it provides of one of the greatest dangers of over-specialisation in production of any manufactured good or primary resource. **Product substitution** can have a devastating impact on the market price and, at worst, destroy an industry. Normally in plantation monoculture, pests and diseases are quoted as the main threat – and their influence should not be underestimated. But the impact of changing tastes and product substitution, in this case by synthetic fibres, should not be ignored. Thirty years ago, sisal used to account for up to 80 per cent of the value of Tanzania's exports. Nowadays, production is on a much smaller scale, concentrated in far fewer plantations, at the cost of thousands of jobs.

Sisal requires well-drained soils and thrives best in hot temperatures. It can withstand three months of drought. Plantations (estates) are usually over 1200 hectares and state owned. A typical 7000 hectare plantation on flat or undulating land may employ up to 2000 men - supervisors, clerks, mechanics, sisal cutters, and weeders. It is divided into sections and worked on a nine-year rotation. On any estate there will be plants at different stages of development to ensure continuous crop production. There will be areas of nurseries, immature and mature plants – with some land left fallow. The sections furthest from the processing centre are cut in the dry season in order to avoid transport difficulties in the muddy summer wet season (December to April). Processing is carried out as soon after cutting the long heavy leaves as possible. When the fibres have been dried, graded, and packed into bales they are exported from ports such as Tanga or Dar-es-Salaam via the well-connected railway network.

Irrigated agriculture has already been illustrated in the production of padi rice. Even in places where rainfall totals are very high, irrigation may be used – simply because of the special requirements of the rice plant. However, it is more usually developed because of a shortage of rainfall combined with high temperatures and so excessive evaporation rates. Many problems associated with agriculture in ELDCs centre on the shortage of rainfall for crop production. Indeed, average rainfall figures may seem adequate enough – but one must not forget that any mean value will conceal considerable variations. Therefore, in any particular year there may only be a 50 per cent chance of obtaining a successful crop – and of course the possibility of frequent crop failure. Even where there is apparently adequate annual rainfall, it may fall at the wrong time of the year – when it is too cold for crop growth or too hot, encouraging evaporation. Another typical problem is rain water lost from highly permeable soil through percolation. Irrigation has an obvious role in the context of these examples.

It is notable that often the most successful irrigation schemes have been based upon a supply of water from streams which rise in well-watered upland areas and then flow across the arid area where water is needed. The Nile is an outstanding example of this kind. Finally, it must be noted that all modern irrigation schemes demand a great amount of capital investment. In ELDCs this capital has usually been provided by EMDCs, although in some cases, notably Iraq, oil revenues have been used to finance projects.

THE ASWAN HIGH DAM PROJECT

The new Aswan High Dam was built eight kilometres south (upstream) of the original Aswan Dam with the aid of American finance and Russian technology. Behind the dam Lake Nasser was formed – a lake as long as England! The original dam could neither produce electricity all year round, nor prevent annual flooding of the Nile - hence construction during the 1960s of this prestige **multi-purpose river project**. Expectations that Lake Nasser could hold sufficient water to supply Egypt's agricultural, domestic, and industrial demands through three years of drought have since been proved. The scale of the project simply cannot be overestimated. Indeed, it has become an additional spectacular tourist attraction of Nile cruises because the previously unreliable water level is now constant and can be navigated all year round. It has thus created a 5000 km waterway to Sudan - so improving trade. Settlement of silt in Lake Nasser has allowed fish to thrive and consequently supports many fishermen. However, less nutrients reach the Nile delta causing poorer fishing in this traditional area. An extra 800 000 hectares of land are now irrigated - pumped electrically to channels for traditional basin, sakia, shaduf, and Archimedian screw methods, but drip and boom systems too. Greater cash crop yields of maize, cane sugar, and cotton have resulted. Also, abundant HEP allows further industrial development - not least at Aswan, *'the Pittsburgh of Egypt'*, where steel, textiles, and chemical fertilisers (to replace the alluvial silt no longer deposited in regular flooding) are produced.

However, no project of this scale can avoid negative environmental and social consequences. As mentioned above, natural siltation downstream has been halted causing retreat of the Nile delta and invasion of previously fertile land by salt Mediterranean water. Indeed, salinisation of poorly drained irrigated land is now reported and an increase in bilharzia spread by snails breeding in the irrigation channels. 50 000 Nubians lost their homes under Lake Nasser and had to be relocated - as did the Abu Simbel temples! Finally, much of the cost is still being paid back - an economic struggle not helped by the demands of Egypt's burgeoning population.

Agriculture in EMDCs

In EMDCs farming is not so much a way of life but a highly specialised **commercial** activity closely linked to trade and industry. Farming regions tend to specialise in the production of particular combinations of products, mostly for sale to urban areas, whether nearby or distant. Improvements in transport, especially the refrigerated lorry, have permitted farms to supply markets thousands of kilometres away. Agriculture employs a small proportion of labour, but availability of capital results in high levels of productivity per worker. Population densities in farming areas are tending to decline as farms are combined into larger units. Such large units, with increased mechanisation, can develop the **economies of scale** central to profitable business. However, the often automatic assumption that EMDC farmers rely entirely on mechanisation and agricultural chemicals may be misleading, given the increasing numbers enjoying considerable success by reverting to **organic** methods.

It is frequently argued that the farming patterns in EMDCs have characteristics similar to those associated with von Thünen's model, especially in large countries where relief is uniform. Broadly concentric farming zones, for example, have been recognised in the USA and Australia. Nearest to major urban markets farming tends to be intensive, indeed highly intensive, with an emphasis on dairying and horticulture. Further away mixed farming is more common. Farming is generally least intensive in areas remote from urban markets, usually consisting of extensive wheat production or cattle and sheep rearing.

Livestock ranching, whether sheep or cattle, earns lower net profit per hectare than any other type of commercial farming - hence its association with environmentally difficult regions. Ranching is an **extensive** system and generally occupies the semi-arid interiors of South America, southern Africa, and Australia - all areas notable for European influence and, consequently, adopted breeds.

LIVESTOCK RANCHING IN THE USA AND AUSTRALIA

This started as an 'open range activity' in the semi-arid grasslands of the USA in the nineteenth century. The land was not divided between different owners, so ownership of livestock was indicated by branding. In these early days hides and wools which could be transported easily were mainly produced. Later, the land was fenced and divided into ranches. Water supplies were provided and railways constructed, thus allowing meat to become the main product.

In general, livestock ranches are larger than any other kind of farming unit in both the USA and Australia. Compared with the amount of land, very little capital or labour is used, but the amount of capital per worker is large - hence labour is very productive. Ranches can be several thousands of square kilometres in size. Their labour requirements are not particularly variable - few people have permanent work. However, at certain times, such as sheep shearing in Australia, large amounts of labour are required. In this case the problem is solved by having teams of sheep shearers travelling from ranch to ranch - paid by the fleece.

In the USA, most of the cattle are European, but also short-horn breeds. In the warmer south, however, crossed breeds with Zebu cattle are used. Irrigation schemes allow the growth of supplementary fodder. Generally speaking, cattle are reared for two years before being sent for fattening on eastern Corn Belt farms.

The interior of South-East Australia sees concentration on Merino sheep rearing, for high quality wool. The poorer quality natural vegetation may support only around one sheep per hectare and water is obtained from boreholes - sometimes artesian. The wool is sent to Sydney, Melbourne, and Adelaide for export.

To conclude - in both these livestock ranching areas relatively little processing of the product is carried out before export. The chief exports are live cattle and wool.

Commercial grain production is now practised on a scale appropriate to supplying world markets - such as North American wheat accounting for much of Europe's bread. Production is currently dominated by the USA and CIS along with Canada and northern China. In the mid-nineteenth century, however, before railways developed, Europe produced all its own wheat, barley, and oats. Late nineteenth century railway development throughout the USA, Canada, South-East Australia, and Argentina allowed the first large-scale production of wheat, which extended during the twentieth century into drier and drier areas. By the early 1930s it had 'jumped' westwards across the USA - a time associated with the 'Dust Bowl' whereby the Mid-Western states suffered appalling wind erosion following, but not necessarily caused by, an unexpected occurrence of several severe droughts. In the mid-twentieth century the former USSR's 'Virgin Land Programme' extended grain cultivation into the dry lands of central Asia (Kazakhstan).

GRAIN PRODUCTION IN NORTH AMERICA

Grain farms specialising in wheat are very large and little labour is used, but due to the high degree of mechanisation, their productivity is high. However, productivity per hectare is limited in this **extensive** system, which may result in relatively low land values.

Physical factors strongly influence wheat cultivation. Level land, for example, is a particular advantage when using large machinery for ploughing and harvesting. Approximately 100 frost-free, sunny days are needed for the wheat to grow and ripen. Where winters are relatively mild, such as in Texas, **winter wheat** is sown in the autumn and ripens the following summer. This usually gives a higher yield per hectare than **spring wheat** which is sown in the spring and grows within 90 days. Winter wheat is, therefore, associated with the south and spring wheat the north -

such as in the vast Prairie Provinces of Canada. Precipitation levels are also influential, with varieties of **hard wheat** (for bread) grown in the driest areas of chernozem soils to the west. Here some farmers use **dry farming** methods, which involve growing a crop in every alternate field so as to store in the fallow soil two successive years' precipitation for its cultivation the following year. Following harvesting, stubble may be left in the fields so as to trap snow which will provide the soil with further moisture. In wetter areas to the east, varieties of **soft wheat** (for biscuits) are grown on prairie soils. The farms are smaller, but more productive because dry farming is unnecessary and so no land is left fallow. Harvesting is often done by hired bands of migrant workers who travel from farm to farm with fleets of combine harvesters. They begin with the winter wheat harvest in Texas (May) and end cutting the spring wheat in Canada (September). Huge elevators beside railway lines store the wheat prior to transport.

Mixed farming, by definition, involves many different kinds of agriculture such as dairying, fruit production, and market gardening. Broadly speaking, it consists of the production of crops in part for sale and partly for feeding livestock, which are then sold. It is notably diversified, highly productive, and often requiring large amounts of both labour and capital. Crop rotations maintain soil fertility and spread the farmer's risk. Little produce is exported – most supports nearby densely populated urban (domestic) markets.

MIXED FARMING IN THE USA CORN BELT

This is based upon the production of maize (corn), but numerous other crops such as soya beans, winter wheat, oats, and alfalfa are included in increasingly sophisticated crop rotations. The area spreading from southern Minnesota, through Iowa, Illinois, and Indiana has low relief, fertile soils, and at least 140 frost-free days per annum. With plentiful rain and a mean summer temperature rising to well over 20°C, the area is highly capitalised and productive. The maize crop is largely fed to animals - mostly beef cattle in the west and pigs in the east (nearer to urban markets). Cattle are moved into the Corn Belt for fattening from the drier ranching areas further west, such as Wyoming. Dairy farming, using maize as a silage crop, increases in importance as you move eastwards across the Corn Belt towards the major urban markets. Throughout the twentieth century, as farm mechanisation has increased, so large corporations have increased their control of tenancies, amalgamating smaller farms for greater efficiency.

Dairy farming is closely associated with urban areas due to milk's bulk and rapid deterioration. The development of refrigerated transport, however, in the 1880s allowed distant countries such as New Zealand to harness climatic advantages to produce processed dairy products for export.

Dairy farming, like market gardening, is **intensive**, requiring almost continual attention by farm workers for the twice-daily milking. Farms are usually small to minimise herding distances and stall feeding is avoided whenever possible to avoid excessive production costs.

DAIRY FARMING IN DENMARK

The glaciated lowland landscape and mild, damp climate of Denmark have been harnessed to considerable agricultural effect. Last century, Denmark was an exporter of grain, but it could not compete when cheaper supplies started arriving from North America. Consequently, it developed dairy farming as an alternative, using high-yield crops and cheap imported grain to feed cattle, pigs, and poultry for the production of milk, cheese, butter, pork, bacon, and eggs for the mass European market. Farms are very small, usually less than 20 hectares, and have long been organised into **cooperatives** for buying and selling. The cooperative creamery, for example, collects milk from the farms in order to produce and market butter, cream, and cheese. Skimmed milk is then returned to the farms to help feed pigs which in turn go to the cooperative bacon factory. The farms, despite being small, enjoy considerable economies of scale, such as in the bulk buying of imported feed or expensive equipment - often financed through loans from cooperative banks. This cooperative system is particularly well respected for the effectiveness of its quality control, advertising, and marketing techniques. These have ensured that 65 per cent of Denmark's dairy produce is exported. Indeed, it ranks in global terms in the export of butter and cheese and produces over half the world's bacon.

Horticulture tends to use great quantities of labour and capital. Resulting productivity per hectare is high and farms are often quite small. Capital investment in heated glasshouses and irrigation systems is not unusual. Like dairying, production is **intensive**, demanding fairly constant attention by the labour force. Agricultural chemicals are frequently applied to ensure high yields of good quality produce. However, vegetables and fruit are often very perishable, necessitating quick delivery to markets. (These may be canning or freezing factories - processing the produce within hours of harvest. Such factories allow the market season of summer producing areas of horticulture to be extended throughout the year.) Irrespective of climate, therefore, great quantities of vegetables and fruit are produced close to urban markets. However, production is also located in areas fairly remote from cities with special climatic advantages.

Vegetables have a much shorter growing season than most types of fruit. Consequently, a wide variety of vegetables can succeed in summer in most parts of Europe, the USA, and southern Canada. But in winter, they can only be produced in the artificial climate of the glasshouse, at great additional expense. Further south, however, several successive vegetable crops such as lettuces, peas, beans, and tomatoes can be produced out of doors throughout the year. Winter production, however, enjoys the greatest comparative advantage, selling at prices high enough to compensate for the transport costs to urban markets of higher latitudes.

Different types of fruit tend to be produced in approximately latitudinal belts, mostly in the northern hemisphere. In cool temperate climates such as in Britain, the main types of fruit grown are apples, pears, cherries, and plums, along with bush fruits such as raspberries. These can all succeed with relatively short, cool summers and are dormant through the winter. In more southerly latitudes, summer temperatures rise and the growing season lengthens. Peaches, apricots, and grapes become important with many of the leading producers around the Mediterranean Sea. Even higher temperatures are needed, as in Israel, for citrus fruits such as lemons and limes, oranges and grapefruits. (The former two cannot tolerate frost at any time.)

FRUIT GROWING IN NORTH AMERICA

North America has sufficient climatic contrasts to produce large quantities of a great variety of fruits – citrus, tree, and bush. Cool temperate fruits such as apples and pears, for example, are grown in western Canada, the Great Lakes area, and New England. Peaches, apricots, and grapes are grown further south. The Central Valley of California, for example, is well known for its grapes and wines, and Georgia, in the east, has specialised in peaches. Citrus fruits are largely confined to southern California, Arizona and the Rio Grande Delta, and Florida, where winter frosts are rare.

13

MINERALS AND ENERGY

Mineral production

Many of the world's minerals are mined from rocks of **cratons** such as the Baltic Shield (Europe), Laurentian Shield (North America), and the Siberian Shield (Asia). These large, relatively stable masses of the earth's crust are also found in the southern hemisphere - related to the ancient continent of *Gondwanaland* - the 'drifted' fragments of which are represented by the Guianan and Brazilian Highlands (South America), Australia's Western Plateau, and much of Africa.

Generally speaking, these cratons in both hemispheres are particularly rich in metallic minerals. Valuable concentrations of minerals also exist in the great ranges of fold mountains that occur along some of their margins.

Global production of some minerals is strongly concentrated in a few places. 75 per cent of gold comes from South Africa, for example, and 66 per cent of uranium from North America.

In a number of other cases two regions produce over half of the world's supply, such as phosphate production dominated by the USA and CIS.

In general, the northern hemisphere is more important for mineral production than the southern hemisphere. (The CIS, USA, and Canada are especially well endowed.) This is because the amount of land is less in the southern hemisphere and the cratons are much smaller. Europe is comparatively unimportant as a mineral producer on a world scale.

Mineral exploitation depends on 3 main factors:

1. The mineral content of the rock (in relation to value).
2. The geological conditions in which mining has to take place.
3. The accessibility of the deposit's location in relation to its markets.

Exploitation is invariably accompanied by extraction of waste rock. If the amount of waste is comparatively great, the mineral is referred to as **low grade**. Low grade deposits are frequently worked if they are near to markets, such as the Lorraine iron field in eastern France. **High grade** ores, in contrast, will be worked in the most isolated locations and difficult environments. The Kiruna iron ores of northern Sweden, for example, have an iron content of around 60 per cent, thus encouraging their development despite the environmental difficulties of this isolated location.

IRON ORE EXTRACTION IN NORTHERN SWEDEN

Location: Very high grade iron ore (50–70 per cent pure) is found at Kiruna and Gallivare in northern Sweden, over 100 km inside the Arctic Circle. Further mining at Svappavaara further east now produces 33 per cent of Sweden's iron. The ore is, however, of relatively lower grade. However, deposits this rich are economic to extract despite the remote, inhospitable environment.

Extraction: The iron ore was originally mined using open cast methods. This is when the surface layers of soil are removed and ore dug out of very large pits. Shift work went on 24 hours a day, even through the winter when there is perpetual darkness and temperatures drop to −40°C. To stop working in such conditions would (just like offshore oil rigs) reduce labour productivity and risk seizure of machinery. Even though mining has now gone underground, continuous working is still maintained.

Transport: The iron ore is moved by rail. The line from Kiruna to Lulea was built in 1892 and the extension, through difficult relief to the Norwegian port of Narvik, in 1902. 66 per cent of the ore is exported through Narvik which, despite its high latitude, is kept ice-free all year by the warming currents of the Gulf Stream. In summer, iron ore can still be exported through Lulea, which being in the Gulf of Bothnia is ice-bound 5 months of the year.

Power: The power used in the mines and on the railway is electricity, coming from an HEP station at Porjus on the River Lule. This is essential as fuels such as diesel would wax at such low temperatures.

Labour: Kiruna has a population of c. 20 000 and everybody depends directly or indirectly on the iron ore mines. Workers are highly paid because of the physical and psychological difficulties of such a location.

Market: The main market is Germany, but some is sent to iron and steel works in southern Sweden. Other minor markets include the UK, such as Scunthorpe via Immingham.

Ease of mining is also of great importance. If the mineral is found at shallow depths, open cast extraction (cheaper than shaft) may take place. Malaysia's dominance in the mining of tin, for example, is undoubtedly related to the relative ease of extraction from alluvial plains. Hydraulic water jets and dredges are used – the heavier tin being separated from the sediments by gravity. It is notable that Malaysia successfully competes in British markets with Cornwall, whose tin has to be extracted from narrow veins in hard granite masses. Not surprisingly, Cornwall's one remaining tin mine constantly dices with closure.

Transport costs are generally unimportant in the production of high value minerals such as gold because values are so high and bulk so small. It is because of this that South Africa can produce so much of the world's gold.

However, minerals of lower value in relation to bulk are much more strongly influenced by transport costs and so production is unlikely to be dominated by a single country. Yet in recent years, the development of bulk carrier ships has enabled the transport of low value minerals over great distances at relatively low cost. Hence iron ore is now transported economically by sea over many thousands of kilometres.

Where large mineral deposits have been discovered in isolated locations, special transport facilities, such as the Kiruna to Narvik railway discussed earlier, have been set up to allow development. Railway maps of both Africa and Australia show repeated examples of routes leading inland from a port, often built especially to transport a mineral ore to purpose-built coastal loading facilities.

MINERAL PRODUCTION TRENDS

Generally speaking there has tended to be a movement of mineral production away from Europe towards the rest of the world. Britain, for example, used to produce half the world's copper in the early nineteenth century. Cornwall used to be extremely important for tin mining. Even as late as 1938 over half the world's bauxite was produced in Europe.

Development of mineral resources outside Europe has frequently begun with the exploitation of precious metals such as silver and gold. The development of cheap, bulk transport, however, has allowed lower value minerals to be exploited. Australia demonstrates this trend well. From the mid-nineteenth century 'gold rush' up to

1940, gold represented half of the value of her mineral exports. Since then, Australia has become the world's leading exporter of bauxite, an important producer of iron ore, lead, and zinc, with nickel, and uranium also being developed. It is notable that many of these developments are in the desert interior.

The 1970s saw a distinct tendency for mineral exploitation to extend from the northern temperate zone into the tropics and southern hemisphere - although the northern hemisphere is still predominant. New mineral discoveries are frequently made in tropical ELDCs such as Nigeria, Zambia, Mexico, and Indonesia. Brazil is of particular note with the discovery of a variety of minerals including uranium, copper, and manganese.

Energy sources

The twentieth century has seen an enormous increase in global energy production and consumption. Early coal dominance has shifted towards petroleum and natural gas. Also HEP has increased in importance and nuclear power has developed. **Alternative sources**, such as tidal, wind, and solar tend to be of local significance.

Much of the world's energy is consumed by a fairly small number of EMDCs where living standards are high. *Per capita* consumption of energy is greatest in North America, the CIS, Australia, and the industrialised countries of northern and central Europe. It is least in ELDCs of central Africa and South-East Asia.

Coal deposits are most prevalent in the northern hemisphere. Relatively little is mined in the southern hemisphere in South Africa and Australia. The great majority of coalfields, producing over 70 per cent of the world's coal, lie in a belt between 30°N and 60°N stretching across North America and Eurasia. Other major producers are China, a number of European countries (where the coalfields are located on the northern flanks of the central European uplands), India, and Japan.

Since coal was dirty, bulky, and difficult to transport, nineteenth century industries tended to cluster on the coalfields, particularly in the older industrial countries of Europe. Now, in these countries and Japan, coal production is declining as competing forms of power have developed. On the other hand, coal production is expanding in the CIS, South Africa, India, and China as heavy industries are developed.

Coal has suffered competition, particularly from oil and natural gas in traditional markets such as railway transport and domestic heating. It is tending to be used now for electricity generation in thermal power stations, the production of coke for metal smelting, and as a raw material in the heavy chemical industry.

Within many countries one can identify a progressive change in the location of coal production according to fairly logical patterns of exploitation. The earliest coal mines extracted the most accessible surface outcrops. These have frequently been worked out and so most mining nowadays is associated with concealed deposits, where plentiful supplies of coal at greater depths are available. In some countries, completely new concealed coalfields have been discovered and exploited in the twentieth century – such as Selby in the UK.

Petroleum and **natural gas** fields are commonly found where there are deep accumulations of sedimentary rocks fairly close to fold mountain ranges. This means that the sedimentary layers are gently folded so as to provide the so-called **trap structure**. Not all trap structures are anticlines of porous rocks with an impervious cap rock above. Fault traps occur, as do traps against rising salt domes near, for example, the Gulf of Mexico.

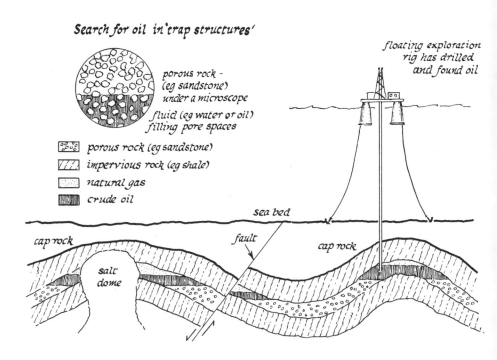

Search for oil in 'trap structures'

porous rock – (eg sandstone) under a microscope

fluid (eg water or oil) filling pore spaces

porous rock (eg sandstone)

impervious rock (eg shale)

natural gas

crude oil

floating exploration rig has drilled and found oil

sea bed

cap rock

salt dome

fault

cap rock

Although the first oil well was drilled in 1859 (in Pennsylvania, USA) it is very much a fuel associated with the twentieth century. Like coal it is a **fossil fuel**, but it cannot be used straight from the ground - it needs refining.

In North and South America, oil and gas fields tend to follow the eastern margins of the Rockies and Andes, both relatively young fold mountains. In Eurasia they occur on each side of the Alpine-Himalayan range of fold mountains and also near the Ural Mountains. The world's greatest petroleum producing area, however, with extremely large reserves, lies in and surrounding the Gulf, chiefly in Saudi Arabia, Iran, Iraq, and Kuwait.

The distribution of the production of natural gas is quite different. Most is produced much closer to the large industrial centres of the temperate zone of the northern hemisphere. Leading producers include the USA, CIS, Canada, and China. Several European countries are also notable such as the Netherlands and UK.

In recent decades the USA has lost its position as the world's leading producer of petroleum and natural gas. The CIS and Middle East have, in contrast, developed greatly. Also, Europe has become increasingly self-sufficient with the development of North Sea reserves.

The chief petroleum producing countries of the Middle East, together with Venezuela, Ecuador, Nigeria, Gabon, and Indonesia are of particular note because it is these producers who form the Organisation of Petroleum Exporting Countries (OPEC). With the advanced industrial countries' increasing dependence upon imported oil in the 1960s, when it was a relatively cheap fuel, OPEC could demand a greater share of the profits of the international oil companies. 1973 saw OPEC exercising considerable 'political muscle' by setting its own price for oil, causing a fourfold increase in less than 2 years! This made previously uneconomic sources viable in environmentally difficult areas such as the North Sea and Alaska. The geography of oil production has changed, and continues to change, as a response to these events.

Attention has turned especially to exploration of the relatively shallow sea bed of the continental shelves where drilling costs are several times greater than on land. Submarine oil fields now contribute a large proportion of the world's supply. Also, development has taken place in remote areas of the world such as the Amazon Basin and the Arctic coast of Alaska (Prudhoe Bay) which, respectively, are linked by pipeline to the Peruvian cost and the ice-free port of Valdez in southern Alaska. Recently, very large oil fields have been discovered in mainland and offshore Mexico. Exploration continues in China, Taiwan, and even the Gobi Desert. Spitsbergen is being surveyed, beneath moving glacier ice, and the Antarctic and South Atlantic could be an exploration bonanza of the future.

Production of natural gas has increased very quickly in recent years. Some natural gas is transported by sea in a liquefied form (LPG) in special tankers but, since most is produced in EMDCs, much is transported direct to consumers by pipeline. Improved pipeline and landfall technology (where the work of coastal geomorphologists is of note) has allowed natural gas to be transported across the bed of the North Sea and the great distance from the Gulf of Mexico to North-East USA.

NORTH SEA NATURAL GAS AND PETROLEUM

The discovery of a large natural gas field at Gröningen in the Netherlands in 1959, and the knowledge of natural gas deposits in North Yorkshire, led to the subsequent prospecting in the southern North Sea. Successful strikes were made in the mid-1960s and especially in the West Sole gas field off the coast of Holderness and in Leman Bank off the Norfolk coast. Most of the gas wells now exist in a broad belt running east-west from Lincolnshire to the Netherlands coast. They are often in shallow water relative to the depths considered fit for oil production. Gas was first piped ashore from the West Sole field to the Easington terminal in 1967. Since then, the importance of natural gas in Britain has increased rapidly. It is of some note that by the mid-1970s it was contributing nearly 30 per cent of Britain's energy supply - not surprising given its considerable advantages for the domestic consumer regarding control and cleanliness. Price regulation has kept it competitive and, undoubtedly, enlarged its market.

Before the development of North Sea petroleum Britain produced very little - mainly from three small oil fields in mainland England. Most of the North Sea oil fields are situated in the northern waters between Scotland and Norway. Here the sea is much deeper than in the gas fields to the south. Severe storms with 30 m waves emphasise the considerable technical achievements associated with exploration and production. A large number of wells have been drilled extending from the Norwegian Ekofisk field, opposite central Scotland, through the large Forties and Piper fields, opposite northern Scotland, to the Brent field off the Shetland Islands and the Magnus field beyond that. Brent is linked by pipeline to the Sullom Voe terminal in Shetland and the Piper field has a pipeline to Flotta in the Orkney Islands. A pipeline from Forties comes ashore at Cruden Bay near Peterhead and continues to the Grangemouth refinery on the Forth. Oil from Ekofisk is piped to Teeside.

The development of North Sea petroleum has undoubtedly been of great economic benefit to Britain, simply because it has removed the need for the country to import very large quantities of oil - although it does import heavier crude than the high quality lighter grades of the North Sea. Britain became self-sufficient during the 1980s allowing a considerable surplus for export, with a notable improvement to the country's balance of payments. Indeed, many people argue that it was only North Sea oil that enabled Britain to 'weather' the economic recession of the early 1980s.

Unlike coal, which can be used directly, oil needs to be transformed into a variety of so-called **refinery products**. Chief among these are petrol and various kinds of fuel oil, including jet aircraft fuel. Oil refineries also supply the raw materials for the petrochemical industries.

An oil refinery may be located either at the oil field (the source of the raw material) or at the market (where the refinery products are consumed). Intermediate trans-shipment points of location are also possible and tend to be associated with coastal location refineries handling imported crude oil.

Oil refinery location has seen a pattern of relocation based on the economies of transport - whether transporting the refinery products to the markets from a raw material location, or transporting the crude oil to the market location refinery. The comparative costs of these two types of transport have strongly influenced the relocation of the oil refining industry.

In the 1930s the great majority of oil refineries were located on or near the oil fields, in particular those of North America and Venezuela, which possessed 75 per cent of the world's refining capacity. Europe possessed very few refineries. Indeed, the reasons for raw material location seem obvious when one considers that only a small part of the crude oil was in fact used, and between 6 per cent and 10 per cent was consumed as refinery fuel. Given the quite small demand for refinery products and the convenient delivery of these products to the consuming areas in small tankers, prohibitive transport costs were avoided by raw material (oil field) locations. Certainly, at this time, North America consumed the greater part of the refinery products it produced, whereas Europe was supplied mainly from Venezuela and the Middle East.

After 1945 the oil refining industry moved strongly towards its markets, particularly towards Europe and Japan, and it has declined relatively, although not absolutely, in North America. The size, location, and nature of the post-war oil market has radically altered. The concentration of demand in the USA prior to 1945 is to be contrasted with the growth of western Europe and Japan, making large-scale refining an economic proposition in these areas. Changes in the nature of demand must be stressed also. Before World War II, the main European demand was for petrol. Since Middle East oil yields only 20 per cent petrol, a great deal of waste would have been transported to market refineries – hence the earlier comment. Since the war, a demand has grown for the full range of petroleum products – diesel, heating oils, heavy oils, and so on. Crude oil is now almost all usable. Hence, although the USA is still the leading oil refining country, Japan, France, Germany, and Italy are also very important despite importing most of their crude oil. It is of note that the UK had become a leading oil refining country well before the North Sea oil fields were discovered.

RELOCATION OF THE OIL REFINING INDUSTRY

1. In the 1930s not all crude oil was able to be changed into refinery products, hence the great loss of weight during refining. Therefore, it was cheaper to transport the 'concentrated' refinery products than to transport crude oil. Nowadays the refining process has become so efficient that there is effectively no loss of weight or bulk at all. Petrol can be produced from any crude, regardless of quality.

2. Since World War II it has been discovered that the transport costs of crude oil can be markedly reduced by the use of pipelines and very large tankers. A 200 000 tonne tanker can run at less than half the cost (per tonne of oil) of a 25 000 tonne tanker. Market refineries can buy from several different fields.

3. Since World War II there has been a great deal of political unrest in oil producing areas – particularly the Middle East. As a consequence, international oil companies have thought it wise to locate new refineries in EMDCs so as to avoid the risk of nationalisation and even confiscation of their assets.

4. In EMDCs, the oil companies can save costs through not having to pay for the provision of an infrastructure of transport facilities, housing, and education institutions. Given the enormous capital costs of building and operating a refinery, it is more sensible to build near your market, rather than near the oil field, which may become exhausted.

5. Some governments in the consumer countries are actively encouraging home (market) location by the application of **preference tariffs** which make it easier and cheaper to import crude oil rather than the finished refinery products. Remember, the development of a refinery industry will help safeguard fuel supplies, save foreign exchange (given that crude oil is cheaper than refined), generate wealth from exporting refinery products, and provide the raw material for the rapidly growing petrochemical industry.

Since the 1970s there have been signs that a further change in the location of the oil refining industry has taken place. Some OPEC countries, such as Kuwait, Libya, and Saudi Arabia, have become extremely wealthy through the rising price of oil and are increasingly using their oil reserves as a basis for the development of a variety of industries based on refinery products. The huge consumption of petroleum products over the world justifies their being transported by very large tankers. Also, the discovery of new oil fields in offshore areas near the UK, Japan, Norway, and other EMDCs means that new raw material sources are being developed near existing refineries. One could argue that the oil refining industry has turned full circle and become raw material orientated again. Current political disunity within the OPEC countries, not least perpetual Gulf instability, overproduction, and consequent collapse in the price of crude oil has vividly illustrated that even this mighty industry is economically vulnerable. Exploration, tanker movements, oil company profits, and so on are consequently reduced whilst the consumers gain the benefits.

Even more recently there has been yet more dramatic evidence to demonstrate just how mercurial the industry can be. The Iraqi invasion of Kuwait in 1990, and resulting UN military response, illustrated once again the significance of petroleum to global geopolitics.

Electricity is an energy form rather than an energy source because it is generated. It also differs from other power sources in that it cannot be stored, although transmission up to 800 km enables it to be very flexible in terms of distribution.

Thermo-electricity uses coal, oil, natural gas, peat, or nuclear reaction to produce steam which powers the electricity generators. Heat may also be obtained from the earth's interior - known as **geothermal power**, such as harnessed at Lardarello in Italy and in California, USA.

Thermo-electricity is providing an increasing share of world production and represents the most rapidly expanding market for the primary fuels. Such thermal power stations are usually able to be located in a country relatively near to large populated areas where there is a great demand for electricity. Coal-fired stations are often near densely populated coalfield areas, oil-fired stations near coastal oil terminals and refineries, and nuclear stations frequently in rural areas on the fringes of large centres of population – thus reducing transmission costs as much as possible given the safety constraints. Within the 800 km maximum cable length, electricity is easily distributed, but the inability to store it makes generation very susceptible to varying demand. Not only are there seasonal peaks of demand, but the daily 24 hour cycle has distinctive peaks and lows. Generating capacity must be capable of matching demand at all times. Thermo-production cannot be switched on and off at a whim, so at less than peak demand the equipment is under-used and uneconomic. Hence such arrangements as the England to France cable, in order to supplement each other's differing peak loads. Likewise the 'off-peak' domestic tariffs in order to encourage use of night storage heaters – or even the pumping of water from lower to upper reservoirs in order to ensure maximum capacity in times of need for pump storage HEP stations such as at Dinorwig, Snowdonia.

Despite the obvious necessity for thermo-electricity production of a ready fuel supply, the principal location factor is a water supply for steam and cooling purposes. A nuclear station, for example, needs 160 million litres per hour, hence coastal locations are common.

HEP stations use the power of falling water to drive generators. They therefore need a sufficient, reliable head and volume of water. Suitable physical and climatic conditions may be great distances from the potential market, hence many potential advantageous locations have not been developed because of prohibitive distribution costs. Generally speaking, HEP transmission costs are very high, yet operating costs remain low because there is no need for any fuel. However, the initial capital costs are very great because of the necessity to build dams to create large reservoirs, and sometimes excavate tunnels so as to collect water from several drainage basins. Indeed, HEP is not the cheapest form of electricity because of the immense installation and high transmission costs.

An alternative type of HEP station uses the regular energy of incoming and outgoing **tides**. The station across the Rance Estuary in France is seen as pilot for other suggested projects such as the Severn Estuary in Britain.

HEP is of particular importance to some EMDCs with mountainous relief, such as Norway, Sweden, Austria, and Switzerland. ELDCs dependent on HEP tend to go for multi-purpose large-scale water control schemes, such as the Aswan High Dam in Egypt, to also control floods and supply irrigation water.

World production of HEP has steadily increased over the last fifty years, but it still only forms a small proportion of the world's electricity supply. In the USA, western Europe, and Japan in particular, it seems unlikely that sufficient suitable sites exist for a major expansion of HEP reservoirs, dams, and power stations. The development of such projects generally meets opposition from people and organisations wishing to conserve the environment for recreational and aesthetic purposes, unless planned with the sensitivity of, say, the Dinorwig pump storage scheme. On the other hand, many HEP projects are being developed in the ELDCs of Africa, South America, and South-East Asia. Indeed, Africa has the greatest HEP potential of all continents.

Nuclear generation increased tenfold globally between the mid-1960s and mid-1970s. The USA assumed dominance, with other leading producers being the UK, Japan, Germany, Canada, Sweden, France, the CIS, and Belgium. To a much greater extent than HEP, therefore, nuclear energy is associated with the advanced, industrial countries of the world. However, ELDCs, such as India, are now developing this contentious power source at a time when many EMDCs are halting expansion programmes in the wake of mounting public scepticism over their cost and environmental and safety records.

14

TRANSPORT AND TRADE

Given that the exchange of goods and services is a central feature of economic activity, not least the development process, a study of transport **networks** (the links that join up points known as **nodes**), **flows** (volume of traffic), and **modes** (methods) is justifiable. One should be familiar with, therefore, the advantages and disadvantages of comparative modes.

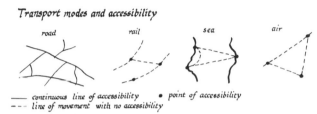

Transport modes and accessibility

road   rail   sea   air

—— *continuous line of accessibility*   • *point of accessibility*
--- *line of movement with no accessibility*

Road transport is easily the most flexible, readily adapting itself to changing conditions such as urban expansion. Railways, however, are only accessible at passenger stations or goods depots. Indeed, they are demanding of space, hence less adaptable than roads. Motorways are, arguably, similar in this respect – demanding much land and only accessible at intersections. However, unlike railways, the same vehicle can travel on both the motorway and the ordinary road system. Inland waterways, including canals, closely resemble railways in the need for an interchange point from barge to road. Sea transport takes this a stage further in specialised break-of-bulk port facilities. Most inaccessible of all, during the journey, is air transport with accessibility restricted solely to airports.

The comparative advantage of one mode over another may be determined by a combination of **direct** line-haul (haulage) costs and **indirect** fixed (terminal and overhead) costs. The latter, for example, will be very high for ocean transport given expensive handling facilities at ports compared to kerbside loading in road transport. However, line-haul costs for road transport rise rapidly with distance because only relatively small loads are carried. Ocean line-haul costs, by contrast, rise comparatively slowly because costs are spread over much larger cargoes. Integration between different modes continually improves with, for example, **containerisation** allowing internationally agreed, standard sized, containers of goods to be transferred between modes without unloading.

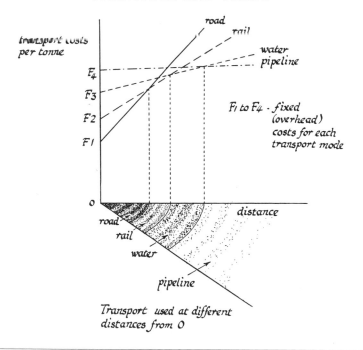

Transport used at different
distances from 0

A MODEL OF NETWORK DEVELOPMENT

The most famous model of network evolution - relating transport to the development process - is that of E.J. Taafe, R.L. Morrill, and P.R. Gould (1963). This six-phase model was based upon studies of Nigeria and Ghana.

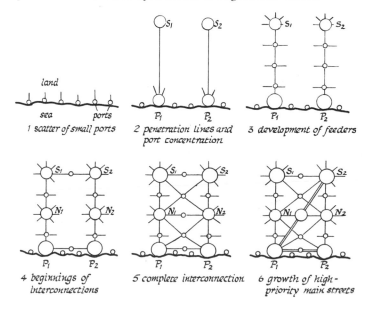

Phase one: A pre-colonial scatter of small ports. Each had limited hinterlands reached by track.

Phase two: Early colonial development of railways to exploit cash crops and minerals. (Two ports, P_1 and P_2, grow and develop links with two interior settlements, S_1 and S_2.)

Phase three: The two major ports enlarge their hinterlands at the expense of smaller ones whilst small centres grow along the main lines of penetration. The economy is dominated by exports.

Phase four: The largest centres link up and intervening centres (N_1 and N_2) start to develop their own hinterlands.

Phase five: Economic activity is at its peak and links develop between all the major settlements. Whilst some feeder tracks are abandoned, national trunk routes are constructed.

Phase six: High priority main links reflect completion of the network.

Phases four, five, and six, therefore, demonstrate increasing connectivity and variety of modes such as metalled roads, rail, and air.

Ghana is a good example because before colonial rule it had numerous coastal fishing ports. British colonial development saw cocoa plantations and railway development to be followed by metalled roads. Construction of the Volta Dam in 1966 allowed navigation for heavy goods, but flooding of north-south routes caused considerable congestion where ferries now have to operate, such as at Yeji. Finally, internal air routes reflect phase six.

International trade can be explained by first looking at the division of labour at an individual level. Individuals work to earn money with which to purchase their various needs from people in other occupations. As a consequence, people become skilful and productive in particular occupations to the benefit of the whole community.

Similarly, the world community will benefit if different countries specialise in the production of certain commodities or the provision of various services. Without international trade each country would have to be self-sufficient and living standards would therefore be lower. (Witness the retardation of progress for so long in the former USSR as it tried, without success, to reach autarky.) Many different types of specialisation can take place – each one giving rise to trade. Climatic opportunities can be exploited for example, such as in South-East Asia where 90 per cent of the world's rubber is produced, and so on.

Large-scale international trade effectively began in the nineteenth century as the Industrial Revolution progressed and various European countries established colonies. Much of the trade involved the importing of food and raw materials in exchange for manufactured goods. Also, enormous areas of newly developed land were opened up in the Americas and Australasia, which became suppliers of food and raw materials for Europe. A further stimulus to trade was the industrial development of the USA and Japan.

Since the mid-twentieth century great changes have occurred. Traditional colonial/mother country trading patterns have declined as the component parts of the European empires have gained independence. Former colonies have frequently become rather small, economically weak members of the so-called Third World. The former colonial pattern of trade has been effectively replaced by trade between EMDCs. Indeed, this represents 65 per cent of the global total, with 25 per cent in ELDCs and only 10 per cent in the former Communist countries.

Since the 1940s there has been a general movement towards removing restrictions to trade, such as the payment of tariffs in respect of imports entering a country. In 1948 the first General Agreement on Tariffs and Trade (GATT) came into operation. Its purpose was to increase the freedom of trade by removing tariffs. Also, other groups of countries joined together to form **common markets** within which trade could flow more freely, for example, the European Free Trade Association (EFTA) and the European Economic Community (EEC) - subsequently shortened to EC and renamed the *European Union (EU)* in 1992. However, such arrangements have not avoided reconstruction, not least because of the dramatic changes in world geopolitics as we entered the 1990s. The majority of centrally planned (Communist) countries have changed to free market (capitalist) economies. For example, the USSR was dissolved at the end of 1991, necessitating the use of hard currency for all trade rather than barter arrangements. Earlier that year, COMECON - the Communist countries' equivalent of EFTA - ceased to exist. 1991 also saw EFTA agree to join the then EC in a 'free market' across all member countries. Effectively, the end of 1992 saw no tariff barriers, or their unseen equivalents, from the Arctic (Iceland) to the Mediterranean (Greece). This European Trade Area (ETA) represents over 370 million consumers! Indeed, given that the EU alone accounts for one-third of all global trade, and will inevitably increase its members in future, the potential economic power within the ETA is considerable. Globally, it is also of note that GATT proved very difficult to renegotiate – hence international trading patterns must always be viewed as changeable with potential for unpredictability.

INTERNATIONAL TRADE TRENDS

In recent years the importance of primary products (food and raw materials) in world trade has declined compared with that of manufactured goods. However, trade in fuels has increased, largely through demand for petroleum. Manufactured products have begun to appear prominently even in the exports of ELDCs such as Bangladesh, Pakistan, and Sierra Leone. Manufactured goods now dominate the export trades of Hong Kong and **newly industrialising countries** (NICs), especially those within the Far East such as South Korea and Taiwan.

Food products see EMDCs as the leading importers, but they also export as much as they import. This is because they consist of two contrasting groups of countries. On the one hand there are the EU countries and Japan, which are net importers – on the other hand are the USA, Canada, Australia, and New Zealand with large areas of farm land and so net exporters of, for example, wheat.

ELDCs are net exporters (although the OPEC countries, many with desert climates, are net importers). India and Sri Lanka, for example, are great exporters of tea, mostly to the UK, while over one-third of the world's coffee exports come from Brazil and Colombia.

Raw materials demonstrate similar patterns as food in that the EMDCs are both the leading importers and exporters. However, they import considerably more than they export. This is because the EU and Japan are such large net importers. On the other hand, the USA, Canada, Australia, and New Zealand are net exporters – the latter two dominating the wool export trade just as the USA dominates the declining cotton trade. These fibres travel mainly to the EU and Japan. Australia and Canada lead in the export of iron ore, mainly to Japan, Germany, and the USA.

ELDCs as a whole are net exporters of raw materials. Over 50 per cent of the world's natural rubber, for example, travels from Malaysia, Singapore, and Indonesia, mainly to the USA.

Fuel is dominated by petroleum and shows a clear distinction between North and South. Over 70 per cent of the fuel exports come from ELDCs, mainly in the form of petroleum from OPEC. Well over 70 per cent of the total imports of fuels go to EMDCs, mostly within the EU, but also large amounts go to the USA and Japan.

Manufactured goods are dominated clearly by exports from the industrialised EMDCs who also, notably, take most of the imports (60 per cent compared to 25 per cent to ELDCs). The EU is both the leading importer and exporter of manufactured goods. The Far East is of particular interest – Japan being primarily an exporter and taking relatively few imports whilst many other, small, Far Eastern countries, such as South Korea and Taiwan, have become prominent exporters. South Korea and Hong Kong, for example, are now leading exporters of clothing. The motor vehicle export trade is dominated by Germany, the USA, and, inevitably, Japan.

Clearly, in conclusion, world trade is dominated by EMDCs.

15

INDUSTRY

Industry refers to all forms of economic activity. Clearly, the resulting variety necessitates classification into categories in such a way as to reflect the position of each activity in a chain leading from resources to market.

FORMS OF ECONOMIC ACTIVITY

Primary industry refers to the initial acquisition of natural resources from the earth's surface or seas.

Secondary industry refers to processing, fabrication, and manufacture of raw materials into finished products.

Tertiary industry refers to administrative and distributive services - effectively linking the first two sectors with the general public.

Quaternary industry refers to information services requiring high levels of expertise, specialisation, and skill.

Alas, there is still not general agreement on the precise differentiation between tertiary and quaternary services. However, if only the first three sectors are distinguished, as in a triangular graph of occupational (employment) structure, assume all services are represented.

When studying industrial location we tend to restrict the term 'industry' to the secondary sector - namely manufacturing. However, there are problems over this definition. The UN, for example, refers to *'the mechanical transformation of inorganic or organic substances into new products, whether the work is performed by power-driven machinery or by hand, whether it is done in a factory or in the worker's home, and whether the products are sold wholesale or retail.'* Clearly, this mouthful is far too all-embracing, hence it is usual to restrict the definition to modern manufacturing industry as associated with specialised labour, powered machinery, and large outputs. Craft or domestic manufacture (including that within the **informal sector**), predominant within ELDCs, is consequently excluded from the definition, necessitating separate consideration.

CLASSIFICATIONS OF INDUSTRY

A number of industrial classification systems have been developed in an attempt to simplify the vast range of industrial activity. It is useful to classify industry in terms of inputs, processes, and outputs - but also its location and ownership.

1. Either **labour** or **capital intensive** depending on the relative importance of each input.

2. The nature of the labour input - whether **unskilled** or **skilled**.

3. The nature of the industrial process - whether **heavy** or **light**.

4. The nature of the output. **Capital** or **producer industries** produce goods which go to other factories to be made into other goods. **Consumer industries** produce goods for direct sale.

5. Location classification - whether **raw material based** or **market orientated**. Or, depending on the relative pull of the major factors of production - whether **tied** or **footloose**.

6. The final classification is based on ownership - whether **private** (individual or corporate) or **state**.

Functional linkages

Functional linkages demonstrate the links that all industries have with each other and the general public, in that no industry is entirely independent or self-sufficient. Links may be inputs or outputs and will vary in complexity, scale, and strength. Generally speaking, local linkages will be more numerous than distant ones and so lead to industrial agglomeration. Links also create chains which bind firms together, reducing costs, but improving efficiency and quality. However, these often complex chains mean that each firm's production is dependent on others. Consequently, problems at one factory will affect all others.

The location of industry

Two complementary approaches may be taken to explain industrial location. A **regional approach** seeks to examine why certain areas are attractive to industry, whilst the **industrial approach** analyses why individual industries are attracted to specific locations. Generally speaking, environmental factors determine the overall regional pattern of industrialisation, whilst economic factors are more likely to dictate individual industries' specific sites. Any final industrial location decision would normally be the product of a complex exercise in weighing up the pros and cons of all potential locations with regard to the specific needs of the firm. The **natural advantages** of an area, such as suitable sites, access to raw materials, and a favourable climate are then reinforced by **acquired advantages** such as good infrastructure, commercial services, skilled labour, market organisation, and the development of subsidiary services. Disadvantages must also be considered – the final decision clearly being where advantages outweigh them.

Factors influencing industrial location

1. **Raw materials** tend to be of far greater significance to early industrialisation. For example, the attraction of a large number of important industries to the coalfields of the British Isles in the nineteenth century as a source of power and raw materials was the result of coal's high transport costs and the steam-engines which demanded large quantities of it. The twentieth century, however, has seen a significant increase in those producer industries which do not use raw materials directly but materials that have been transformed to a greater or lesser degree, such as components in the car or electronics industry. The weight of one material alone has, therefore, been decreased. In addition, there has been an increasing availability of more mobile sources of power, a relative decrease in the cost of transporting raw materials compared to products, and a greater efficiency in the use of raw materials, thereby reducing the weight of materials used and increasing the possibilities of using what was formerly waste as by-products. Finally, developments in transport and an increasing density of transport networks in EMDCs have reduced problems of supply.

However, for some industries, proximity to raw materials is still important – particularly for those with a high **material index**.

$$\text{Material index} = \frac{\textbf{weight of raw material used}}{\textbf{weight of final product}}$$

The higher the index, the more an industry will be attracted to raw material locations. This is because the industry, if not located at the raw materials, will find itself paying transport costs on waste, assuming no by-products can be made from it.

In general terms, therefore, industries that lose either great bulk or weight in the manufacturing process, or which use perishable primary products, such as fruit for fruit canning, will try and locate near the raw materials because of the high costs of transporting waste. Examples include copper, iron and steel smelting, and cement manufacture.

The tie of industries to raw materials also depends on the value of the raw material in relation to its weight. Low value, high weight raw materials attract industries to their source - such as copper smelting. Also, industries in which the raw material makes up a very high proportion of the final product's cost find a raw material location advantageous.

2. **Power supplies**, historically coal, have been a critical factor in the location of industry. Inertia (see later), however, dictates that much industry remains on the coalfields. This is because of economies of scale in the provision of services resulting from industrial agglomeration, the development of high population concentrations, and consequently markets, and the existence of a large amount of fixed and immobile capital investment - all giving the coalfield areas considerable forces of attraction. Power, like raw materials, is nevertheless declining in importance as a location factor nowadays. Fuel efficiency has improved considerably. For example, a fraction of the amount of coal needed to smelt iron in the eighteenth century is needed today. Also, the advent of more mobile sources of energy such as oil, natural gas, and, especially, electricity has increased the number of potential locations. Electricity grids and oil and gas pipelines have undoubtedly made modern growth industries less locationally demanding - footloose rather than tied.

However, for some industries proximity to cheap fuel or power supplies is critical. Some industries, for example, require vast amounts of cheap electricity. Electro-metallurgical and electro-chemical industries are often located near HEP stations as a result - such as aluminium extraction from bauxite in Kitimat, Canada.

3. **Transport** is an essential consideration for industry. Consequently, areas with efficient transport networks will attract industry. The reduction of transport costs (more accurately expressed as **transfer costs**) is central to the problem of industrial location. Transfer costs are largely made up of the direct line-haul costs of transporting goods from place to place, but also indirect costs such as depot overheads, insurance, damage in transit, clerical,

and administrative costs. Hence, the reduction of distance is critical – not just the length but the nature of the journey. Therefore we talk of **economic distance** which is affected by other considerations – namely the mode of transport, the type of commodity, freight rates, and so on.

A basic concern of any manufacturer is to reduce economic distance. Consequently, transport plays an important role in industrial location.

4. **Labour costs** and **availability** are never easy to ascertain because true costs are not just about wages and salaries but the whole question of productivity. Wages have to be balanced against output – a balance very much affected by labour relations, absenteeism, and rates of turnover (all factors which vary spatially). Quality and quantity of labour bring in the question of skills – certain areas offer certain skills whilst, overall, automation has now reduced the need for many crafts. Indeed, the increasing significance of part-time and female workers (the latter with proven patience and dexterity in modern assembly processes) has much affected traditional perspectives on industrial labour. Generally speaking, however, an area with a skilled and versatile labour force will be attractive. Yet another consideration is the mobility of labour – whether between jobs or areas. If a workforce were completely mobile, labour would have no locational pull because workers could move to jobs. But labour is relatively immobile, with only young, professional, and skilled workers showing increasing mobility. An immobile labour force will, therefore, attract industry. Wage rates negotiated nationally may be less apparent of late, but regional differences are often minor and so locationally insignificant. (However, weighted allowances, 'golden hellos', and so on are increasingly notable in, for example, South-East England where the cost of living remains higher.) Internationally, wage rates are of critical significance, particularly when low labour cost examples are quoted from the Far East and trans-national company **globalisation** of production becomes commonplace.

Finally, management is especially important in that industrial success depends on their entrepreneurial enterprise and initiative. However, management skills are by no means evenly distributed. For example, one of the major obstacles to industrialisation in ELDCs is the shortage of management skills. An area will only be attractive to industry if it is likely to produce good managers or attract them from elsewhere. Hence the pull of South-East England which possesses large numbers of high-calibre executives – and the leisure, recreational, and cultural opportunities likely to attract others.

5. **Capital**, whether fixed (land, buildings, machinery), or working (stocks, products, money) differs in its effects on industrial location. To acquire **fixed capital** will cost different amounts in different places. Obviously, areas where costs are lower will be more attractive. Certain industries – especially small or new firms – may benefit from renting or buying and converting existing properties. For example, the textiles industry which can use multi-storey buildings or electrical engineering. However, it should be noted that fixed capital is relatively immobile and variations in the price of land can have important effects on industrial location and relocation. Heavy investment in fixed capital means that it is often cheaper to expand production in an existing location rather than move and build afresh elsewhere. This is because land, buildings, and large machinery are relatively immobile. Hence **geographical inertia**[1] is commonplace – arguably an industrial paradox yet reflecting the rational judgement of industrial decision-makers.

The availability and cost of **working capital** also varies regionally – hence major investment centres such as London becoming industrial centres. International variations are of particular note. ELDCs, for example, particularly with unstable or unpredictable political regimes, may find it difficult to attract investment capital. Security and the certainty of returns are important considerations for the lender. This is particularly pertinent nowadays, given the **debt crisis** and repayment defaulting associated with many ELDCs and NICs, notably Brazil and Mexico. Indeed, many 'western' banks 'wrote off' ELDC loans in 1986 and 1987. Large industrial concerns such as trans-national corporations attract investment with far greater ease than small-scale, independent industries. Likewise, expanding, large, successful regions within countries, such as South-East England within Britain, attract investment more easily than relatively depressed areas.

6. **Markets** nowadays influence greatly, if not dictate, most industrial location decisions. It should never be forgotten that the purpose of industry is to produce goods to sell at a profit – hence you cannot ignore the market as a locational factor. Indeed, as the pull of raw materials and power has declined, the market has become the prime consideration, especially as there is now a general tendency (with bulk raw materials

---

[1] Geographical inertia occurs when the original reasons for their location no longer apply, yet firms remain in the same locality.

transport) towards increased transfer costs on finished products. This is particularly so in industries which produce low unit value products or products which increase, rather than decrease, in weight or bulk by processing, such as furniture and brewing. Also perishable and/or fragile products need a market location. The market can even be the source of raw materials, such as scrap for the steel industry. Certainly, assembly industries may be regarded as markets for components. Indeed, **just-in-time** delivery systems of these (ensuring quality control) are possible, so facilitating **lean production** methods. The market is very much a concentration of labour and spending power, hence the larger it is, the greater the attraction it exerts. In ELDCs the limited size (through low income potential) of markets is recognised as a limiting factor to industrialisation. Even in EMDCs, regional variation in sales potential can encourage concentration of industry in already prosperous, successful areas. A large and expanding market permits economies of scale, encourages the development of specialised services, such as banking, insurance, repair and maintenance, and so on, permits personal contact between supplier and consumer, and often leads to more favourable transport rates because the generation of traffic is greater. The large and varied pool of labour is another notable advantage - particularly female labour for precision industries such as electronics.

However, market areas do suffer from **diseconomies of agglomeration**. Competition for labour can push up wage rates, land for expansion becomes scarce, and so increases in price. House prices inflate and traffic congestion reduces accessibility and increases delivery time.

7. **Government intervention** in industrial location is a most important factor to consider. Although manufacturers will try to locate where they think they will get the best returns, their choice is by no means entirely free because governments are normally concerned to encourage, or restrict, industrial developments in certain areas for various reasons - whether economic, social, political, or strategic. Intervention should not be assumed to be a modern historical relic of the former centrally planned (Communist) economies of eastern Europe. It has been and remains common within free market economies such as Britain's.

GOVERNMENT INTERVENTION IN BRITAIN

The Industrial Revolution saw Britain's agricultural economy changing to an industrial one. Northern cities grew and prospered in coalfield areas, whilst southern regions (excepting the capital - London) declined. However, the twentieth century has seen a marked reversal of fortunes, with southern prosperity based upon a more diversified light manufacturing industrial base with increased emphasis on service occupations, such as finance. Northern regions, by contrast, have declined through **deindustrialisation**. No longer the *'workshop of the world'*, their heavy industries were increasingly less equipped to compete with foreign competitors unburdened by high wages, old-fashioned working practices, and outdated plants.

Ever since the Great Depression of the late 1920s to early 1930s, governments of all political parties - labour, conservative, even coalition during World War II - have addressed the socio-economic and environmental problems resulting from northern deindustrialisation and population *'drift to the south'*. The general theme has been how to encourage new industrial development in the North whilst preventing the South from becoming too congested. Up to the Thatcher administrations of the 1980s, variations on the 'stick and carrot' approach have been operated. The **stick** was planning controls in the South. Statutory Industrial Development Certificates (IDCs) and Office Development Permits (ODPs) were often refused, but offered in the North. The **carrots** were all sorts of incentives in declining northern regions designated as Assisted Areas (called various names by different governments and aided according to the political pendulum). Incentives included tax allowances, low interest loans, grants, subsidies, even ready-built factories. The 1980s, however, saw a distinct change in emphasis. The stick was progressively removed, with evential abolition of the IDC and ODP, so as not to hinder further development in the South. The carrots were made far more selective, with fewer and smaller Assisted Areas and especial emphasis on individual centres with acute problems, designated Enterprise Zones.

The success or otherwise of the **regional assistance** central to government intervention in industrial location remains the subject of perpetual political debate. However, objective examples of success can be stated, such as Ford's Halewood plant in Liverpool, producing the best selling *Escort* since the 1970s. Another example, and far more recent, is the creation of 3000 new jobs in Bridgend, South Wales. Again Ford were involved, having been offered significant incentives to invest £60-£70 million in expanding their engine plant to produce 'lean burn' (lead free) engines for their mid-range cars. Finally rationalisation of steel production in Scunthorpe during the 1980s (resulting in 35 per cent male unemployment) led to Enterprise Zone status. This recently expired following a return to relative prosperity based upon the more diversified industrial base attracted.

8. **Water supplies** are of increasing importance, but not a major factor. Water is used for processing, cleaning, raising steam, cooling, and waste disposal. Both reliability and quality of supply needs to be considered, and the quantities never underestimated. For example, it takes 200 litres to produce a pint of beer – 450 000 litres for a car! As the demand for water increases, so do the problems of **waste disposal**. Indeed, sites with expensive disposal problems are not going to be attractive to industry, especially with increasingly strict pollution control legislation within most EMDCs.

9. **Chance** or **luck** may well determine an industrial location with no apparent economically logical reason in support. The classic example for this is the huge Rover Group (previously Morris) car plant at Cowley in Oxford. This started as a bicycle repair shop in the converted school of William Morris' father!

Industrial location theories

The factors affecting the location of manufacturing industry in the real world are complex and interrelated. It is for this reason that academics have attempted to simplify real world situations by isolating what they consider to be the key factors in explaining location.

Resulting theories and models fall into two main groups – those seeking the least cost (optimum) location as investigated by A. Weber and A. Lösch, and those seeking areas of profitability (satisfactory locations) such as studied by D.M. Smith.

A. WEBER'S INDUSTRIAL LOCATION THEORY (1909)

Weber aimed to explain industrial location in terms of three economic factors. These were transport costs, labour costs, and agglomeration economies. Of the three, transport was the most important (and included raw material and power costs).

Agglomeration economies are the economies of scale derived by firms clustering together and sharing ancillary services and public utilities. They also include industrial linkages, the development of a specialised labour force, bulk buying, and shared marketing. **Deglomeration economies** involve the weakening of agglomeration economies by dispersal of firms due, for example, to land price inflation.

Weber's theory necessitated a number of simplifying assumptions in order to simplify reality. The similarity of many of these to those in von Thünen's model of agricultural land-use (discussed in *Agriculture*) and W. Christaller's central place theory (see *Rural Settlement*) should be noted. Landscape was envisaged as a flat, featureless (isotropic) plain across which there was an uneven distribution of resources. Thus the raw materials, fuel, and water needed could only be found in given locations. The size and location of markets were stated, as were several fixed locations of labour where given wage rates operated. Labour was immobile and unlimited at these locations. The plain had a uniform culture, race, climate, political, and economic system. Perfect competition was assumed whereby resources and markets were unlimited, at their given location, and no firm could become a monopoly. Entrepreneurs sought to minimise the total cost of production. The cost of land, building, equipment, interest, and depreciation of fixed capital did not vary regionally and there was a uniform system of transport.

Transport costs depended on the classification or raw materials as follows:

Localised – **pure** (lose no weight in processing),
           gross (do lose weight in processing).
Ubiquitous – available anywhere, such as water.

In addition, the weight of the manufactured product was considered by the construction of the material index discussed earlier.

$$\textbf{Material index} = \frac{\textbf{total weight of the (localised) raw material}}{\textbf{weight of finished product}}$$

Industries using pure raw materials will have an index of 1 as there is no weight loss in production. An index greater than 1 indicates a weight loss. Generally, the higher the index, the greater the weight loss and so the closer it will be to the raw material location. Industries with an index close to or equal to 1 will have a market location since the cost of transporting the finished product is much greater than the cost of transporting any one of the pure raw materials from source.

In Weber's locational triangle, RM_1 and RM_2 are two localised raw materials and the market is at M. X is at the mid-point of one of the triangle's sides. 1000 tonnes of each raw material is required to produce the product. With both raw materials pure the market would represent the least cost location. With both raw materials gross, losing 50 per cent of their weight in manufacture, X becomes the least cost location. However, if RM_1 is pure and RM_2 is gross, losing 50 per cent of its weight in manufacture, RM_2 becomes the optimum site.

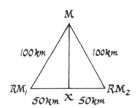

Weber represented transport costs by **isotims** (lines joining points of equal transport costs) increasing regularly with distance.

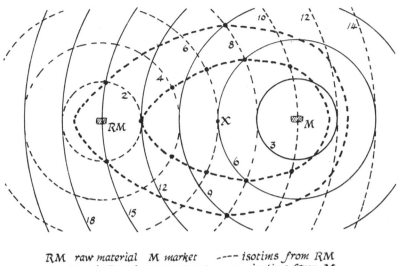

RM *raw material* M *market*   ---- *isotims from RM*
---- *14 and 17 total transport cost* ——— *isotims from M*
isodapanes

Isodapanes are lines joining points of equal total transport costs. Once plotted, they can be read like contour lines showing a basin to reveal the least cost location.

The deflecting effect of labour costs

A particularly productive or efficient labour force at X could deflect the industry there if the savings in labour costs equalled or were greater than the additional costs of transport involved in moving from M. Total transport costs at X are 12 units and assuming savings on labour costs (at X) to be 4 (transport costs) units, then this would be greater than the additional costs of transport (2 units) and so it would be sensible to move to X.

To simplify this procedure the isodapane with the value of the labour savings is identified as the **critical isodapane**. If the site of cheap labour is located within it, then it is profitable to move. If it is outside the critical isodapane, then it is not.

Labour savings for example at L_1 and L_2 are equivalent to 3 transport cost units. You could profitably move to L_1 since the savings on labour exceed the transport costs – it is within the critical isodapane. L_2 is outside it, hence not a sensible move.

L_1 L_2 *points of cheap labour*
X *least-transport-cost location*

isodapanes show the increase in transport costs if production moves from X

critical isodapane

The deflecting effect of agglomeration economies

As with labour costs earlier, savings through agglomeration may prove sufficient to deflect an industry from the least transport cost location.

Again critical isodapanes representative of the savings are identified and if they intersect it would be economic to relocate within the zone of intersection.

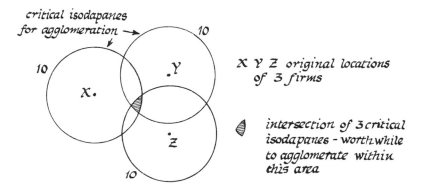

critical isodapanes
for agglomeration → 10
10

X Y Z original locations
of 3 firms

intersection of 3 critical
isodapanes - worthwhile
to agglomerate within
this area

Appraising Weber's theory

Criticisms focus on two aspects of the work - that it is now rather dated and that the fundamental assumptions are too simplistic. Certainly it underestimates twentieth century technical revolutions in transport and raw material processing - not least the growth of light industries increasingly divorced from direct processing. Also there has been far more government intervention than Weber could envisage. The assumption of perfect competition, with no changes in demand or price from set markets, is hardly credible. Markets in reality are dynamic areas of varying size served by elaborate transport networks and modes. Finally decision-makers are not always rational. There is much, therefore, to question - not least misconceptions in the original assumptions. Despite the acknowledged validity of many of his conclusions, Weber's theory is too abstract - more of academic than practical value.

Illustrations adapted from Bradford and Kent (1977)

This search for best location is seen, therefore, as too unrealistic in that firms normally operate within areas of profitability, the extent of which will depend on variations in cost and price, as determined by demand. Finding the margins of these areas and explaining the **sub-optimal location decisions** within them is seen as more realistic. This behavioural approach, with the decision-maker as a **satisficer** rather than **economic man**, is best demonstrated in the work of D.M. Smith.

THE SPATIAL MARGINS TO PROFITABILITY
(D.M. Smith, 1971)

This model considers the interaction between costs and revenues in a spatial context. So long as costs do not exceed revenues, profits can be made. By using cross-section diagrams Smith could illustrate various circumstances such as constant demand hence revenue, constant production costs, or variations in both. Not only are the **spatial margins to profitability** shown, but the extent of profitable areas as determined by the steepness of the **space-cost curves**. Thus, whether firms have to be concentrated or can, in contrast, be dispersed is demonstrated. Clearly, however, whilst one firm locates at the optimum site of maximum profit, all others have sub-optimal locations.

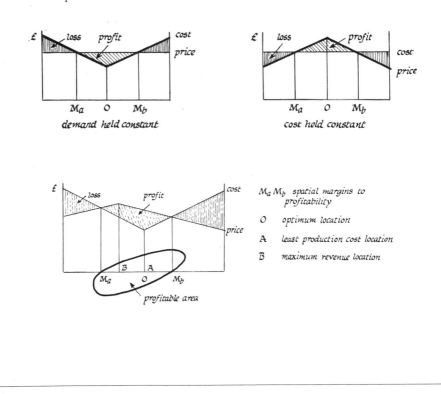

$M_a M_b$ spatial margins to profitability

O   optimum location

A   least production cost location

B   maximum revenue location

Attention in recent years has turned to **how** and **why** firms choose industrial locations. In reality it is very difficult to find the optimal location. Indeed, some companies are more successful than others in choosing sub-optimal ones.

A.R. Pred's (1967) **behavioural matrix** is useful in this context. Note how the biggest profit makers enjoy the best quality and quantity of information whilst having the greatest ability to use it. Consider in contrast, the circumstances of the loss makers.

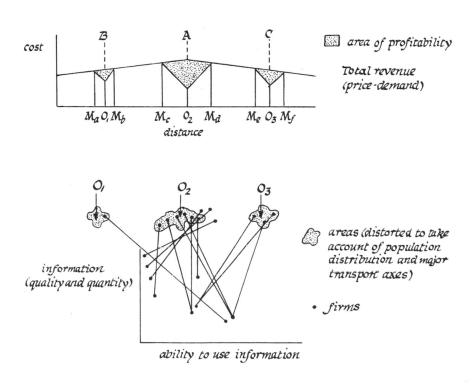

Illustration after P.J. McBride (1980)

In conclusion, the real world is, to say the least, complex and industrial location is a demanding subject. However, there is some order in this complexity and the geographer's role is to try and simplify reality and so allow the order to be demonstrated as a prerequisite to understanding. Clearly, the early stress on seeking optimum location was not realistic, hence the change of emphasis towards finding areas of profitability and explaining deviations from the optimum. Indeed, research tends to emphasise that finding new locations for factories rarely happens. Most companies modernise and/or expand their existing premises. Even more revealing is that when new locations are sought, cost evaluations follow, rather than precede, the decision. Research emphasis on the behaviour of decision-makers is, therefore, justified.

The distribution of industry

World-wide agglomerations of industry at different scales are clearly apparent. A so-called 'power belt' extends from the Mississippi to the Ural Mountains, accounting for around 80 per cent of the world's total value of manufactured output and 90 per cent of its energy consumption. Major concentrations of industrial development dominate in North-East USA, Europe, western CIS, northern India, and East Asia. Minor concentrations are also notable, as are the large areas devoid of industry. Generally speaking, most industrial development is located in the middle temperate latitudes and is significantly associated with coalfields and areas of European settlement. This is in marked contrast to tropical latitudes, characterised by relatively little.

Industrial regions

In the now advanced industrial regions, early (Industrial Revolution) growth was associated with the exploitation and use of coal power, accumulation and investment of capital in industry, and growth of a large and varied labour force. Urbanisation, the development of transport networks, and a complementary Agricultural Revolution were, therefore, necessary accompaniments. Geographical inertia subsequently maintains the coalfield association as infrastructure developments, industrial linkages, the immobility of heavy capital investment, and the expanding market dictate. Elsewhere, **entrepot**[1] locations, such as North-East USA, evolved and continue to thrive as a result of ocean trade. The contemporary increasing tendency towards market orientation of industry and services, discussed earlier, is explained in summary by more mobile energy sources, technical changes in raw material processing, and market concentration of capital and labour.

Also there is increasing industrial relocation in the less expensive, less congested urban-rural fringe with heavy industries gravitating towards coastal break-of-bulk points.

In ELDCs there are far fewer concentrations of industry. Indeed, much industrial activity is still concerned with primary processing of raw

---

[1] Entrepot - a centre to which goods in transit are brought for temporary storage and re-exportation.

materials for export. Certainly, where manufacturing is better established, often involving trans-national corporations, limited infrastructure (especially communications), poor quality labour (in terms of health and education), and a poverty-restricted domestic market potential are frequently cited hindrances.

Economic growth and development

To those within EMDCs, economic development is synonymous with wealth as measured by gross national product (GNP) *per capita*. (This is the value of all goods and services produced, divided by the country's population – a value including 'invisibles' such as overseas financial services not considered in the gross domestic product.) However, purely material standards of living are increasingly perceived as too limiting as an expression of development. Hence, social, cultural, and welfare criteria, expressing quality of life, are used in addition – such as the **physical quality of life index** (PQLI). This averages literacy, life expectancy, and infant mortality on scales of 0–100 (0 the 'worst' country, 100 the 'best'). Clearly, such quantification is problematic, not least because of the plethora of other, interrelated factors which influence quality of life and contribute further to any appreciation of comparative levels of development. Furthermore, studies of correlation between independent variables, such as GNP *per capita*, and dependent ones, such as infant mortality, industrial energy consumption, illiteracy, and number of persons per doctor, add to our understanding – whether assessed visually on scattergraphs, or tested statistically by the Spearman rank correlation coefficient test.

SPEARMAN RANK CORRELATION COEFFICIENT TEST

This is a non-parametric measure of the relationship between two sets of ranked variables. It therefore makes no assumptions about the population and takes us much further than judging visual correlation – such as by comparing choropleth maps or plotting a line of best fit on a scattergraph. The test not only tells us if the variables correlate but how well they do.

The relationship between GNP and steel consumption (both *per capita*) might be expected to reflect comparative levels of economic development – the higher the wealth, the higher the level of development and the greater consumption of steel – and *vice versa*. The validity of the initial relationship may be statistically determined by this test.

	GNP per capita $	Rank order GNP	Steel use per capita kg	Rank order Steel	d	d^2
Chile	1310	7	65	9	-2	4
Egypt	680	9	70	8	1	1
Germany	14400	3	530	2	1	1
India	300	10	20	10	0	0
Italy	10350	4	450	3	1	1
Japan	15800	2	550	1	1	1
Portugal	2830	6	140	6	0	0
Turkey	1210	8	105	7	1	1
USA	18500	1	410	4	-3	9
Venezuela	3230	5	190	5	0	0

$$\Sigma d^2 = 18$$

$$r_s = 1 - \frac{6 \times \Sigma d^2}{n(n^2 - 1)} = 1 - \frac{6 \times 18}{10(100 - 1)} = 1 - \frac{108}{990} = 1 - 0.109 = 0.891$$

where
- r_s = Spearman's coefficient
- d = the difference between the ranks
- Σ = total or sum of
- n = number of variables

r_s can have a value ranging between -1.0 indicating a perfect negative correlation and +1.0 which is a perfect positive one. A value of 0.0 indicates no relationship at all - therefore the nearer to minus or plus 1.0 the stronger the correlation. Furthermore, published **critical values** of r_s for stated numbers of pairs (**n**) allows one to state the statistical significance of the result.

	Confidence Level	
	95%	99%
n = 10	0.564	0.745

In geography a 95 per cent confidence level is adequate. 99 per cent confidence is, however, desirable because one can state with certainty that there is only a 1 per cent chance of error. In this example r_s = 0.891 exceeds 0.745 and so one can state with 99 per cent certainty that there is a positive correlation between GNP and steel consumption (both *per capita*) and so the original relationship has been proved.

Wealth inequalities, occupational structures, dietary variations, health care, and educational expectations – not least gender roles – characterise relative levels of development. Indeed, global inequalities in such variables appear to be widening. However, as suggested above, simplistic assumptions relating quality of life solely to economic development must be avoided, if only because 'development' is, arguably, an indeterminate concept.

Just as precision in defining economic development eludes us, so does explanation of the development process. Many theories have been formulated, with varying degrees of emphasis placed on, for example, the organisation of labour, resource development, and technical innovation.

W.W. ROSTOW'S MODEL OF ECONOMIC GROWTH (1960)

This 'western' model has been very influential, yet it lacks any mechanism for linking each stage from subsistence to an advanced economy.

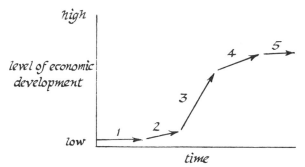

1. the traditional society 2. preconditions for take-off 3. take-off 4. drive to maturity 5. age of high mass consumption

Stage one: Subsistence agriculture dominates. Natural resource potential, rather than realised asset.

Stage two: Extractive industries start; ideas and inventions improve agriculture; economic growth speeds up as trade expands.

Stage three: Old traditions are overcome and the modern industrial society is born. Manufacturing industry grows as investment increases. Growth becomes self-sustaining.

Stage four: The industrialisation is consolidated – strong companies thriving, weak failing. The economic growth spreads to all areas and infrastructure improvements continue.

Stage five: Society becomes more materialistic through advertising, but social welfare is also of note.

Whether or not this can be applied to ELDCs remains the subject of debate. It is notable, however, that external aid is often cited as a means of speeding up the process to point of 'take-off' beyond which no more help should, in theory, be needed.

G. MYRDAL'S MODEL OF CUMULATIVE CAUSATION (1956)

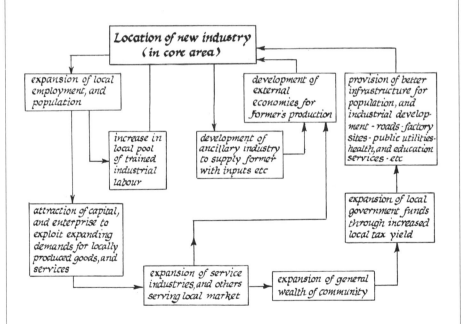

This model effectively suggests unbalanced growth through economic development concentrated in **core areas (growth poles)**. This development would initiate others in a **multiplier effect** known as **cumulative causation**. Clearly, regional differences would intensify by operation of **backwash effects** whereby raw materials, skilled labour, investment capital, and so on are drawn to the core from the surrounding periphery. These backwash effects are further amplified by the core flooding the periphery with more competitive goods, so hastening its decline.

However, **spread effects** should eventually operate as demand from the core requires additional products from the periphery, so stimulating industrial and infrastructural developments there. Alas, this is rare in ELDCs which lack the governmental financial resources to stimulate spread effects by regional policies to invest in peripheral infrastructural developments or provide investment grants to new industries. Even in EMDCs **distance decay** is frequently apparent, in that peripheral areas nearest core regions enjoy the strongest spread effects, whilst remote regions remain virtually unaffected.

E.F. Schumacher's alternative course for development through intermediate technology (1973)

Increasingly, contemporary development planning follows the visionary approach of Schumacher's Intermediate Technology Development Group (ITDG). This is because capital intensive, labour-saving 'western' technology, in circumstances of scarce finance and abundant labour, is proved too often to be inappropriate. **Intermediate technology** involves small-scale local industrialisation in rural areas using inexpensive, labour-intensive equipment within easy reach and comprehension of the poor. Examples abound, such as bicycle wheels used for textile spinning in Sri Lanka. However, intermediate technology's low productivity and lack of prestige does not always endear it to the governments of ELDCs eager for glorifying projects, especially if funded through foreign aid. But given the reality of contemporary ELDC occupational structures, with 60 per cent or more seeking work in the informal sector, the ITDG approach seems particularly appropriate.

16
RURAL SETTLEMENT

Settlements represent the most visible expression of human culture on the natural world. Their development may be considered as our principal adaptation of the environment in order to suit our own needs. Settlement geography is, consequently, a key component of the discipline allowing understanding of site and situation, pattern and distribution, development, structure (morphology or form) and zone of influence.

The study of settlements may be approached in two ways – either examining them in isolation as separate entities, or as integral components of the environment. However, prior to any examination, classification is essential in order that valid comparisons may be made.

Each settlement is, of course, unique – but they do fall naturally into groups. For example, classification may be as temporary or permanent or by size, age, function, or structure. Even site, situation, cultural characteristics, and building materials may be used in classification. However, the twofold distinction between rural and urban categories, although imprecise, is the most basic distinction.

Definition of **rural** is not straightforward. For example, *'non urban'* is too vague – *'agricultural'* too presumptuous. However, ranging in size from isolated dwellings to villages they are, undoubtedly, an essential component of the human landscape and account for just over half the world's population. Indeed, agricultural settlements were the world's first stable communities – socially cohesive and defensible expressions of the socio-economic progression from hunter-gatherer to sedentary farmer. For example, we can trace the first villages back to the Neolithic Age (6000–5000 BC) in Mesopotamia, Egypt, and the Lower Indus Valley.

Britain's history of rural settlement goes back to the Celts who cultivated exposed upland sites 4000 years ago – hill fort villages long since abandoned for more sheltered locations. Indeed, the so-called Dark Ages saw the densely wooded English lowlands colonised first by Saxons and later by Danes and Normans. This multiplication of villages meant that by the time of the *Domesday Survey* of 1086 most of today's settlements were already in existence. Medieval England saw settlement sizes increasing. However, this overall consolidation of village life was interrupted by the Black Death (1348–1350) which killed one-third of the English population and led to the desertion of 1300 villages identified nowadays only as isolated derelict churches or faint ridges in the ground. Finally, during the Industrial Revolution (late eighteenth to early twentieth century) it is notable that non-agricultural mining and quarrying rural settlements developed.

Studying rural settlements involves both primary and secondary data collection. Fieldwork, including the use of historical and contemporary maps may elucidate origins, historical development, and present functions. Place names are a particularly useful tool - for in Britain they incorporate both meaning and cultural origin.

AN EXAMPLE OF 'ONOMASTICS' (PLACE NAME ANALYSIS)

Brigg on the River Ancholme in South Humberside is shortened from Glanford Brigg. 'Glanford' can be traced back to the Saxon word 'gleam' meaning merriments. Glanford therefore means the river fording point where sports were held. The later addition of Brigg acknowledges the eventual bridging of the river.

Types of rural settlements

Classification solely by size, function, or form will invariably lead to problems due to the many exceptions. For example, primary functions such as agriculture or mineral extraction may well be likely but certainly not exclusively so given widespread commuting from **dormitory settlements**. Likewise the *'town by size yet village by character'* and *vice versa* issue is also inevitable if a population of 2000 is used as the crude upper limit before rural becomes urban. For example, the small market town of Brigg, mentioned earlier, is particularly well endowed with varied functions and services, yet its population of around 5500 is only marginally greater than the very poorly served neighbouring 'village' of Broughton. The criteria, therefore, for classifying rural settlements are both subtle and varied. Indeed, one should not dismiss the value of instinctive 'feel' and certainly listen receptively to the residents' own perceptions when classifying any settlement.

For example, **isolated dwellings** such as farms and manor houses are likely to be effectively self-sufficient expressions of economic rather than social need. **Hamlets** may have grown around them - the small group of houses perhaps served by a church. It is only at the level of **small village** that a wider range of service functions would be evident. For example, the church would no longer be the only centre for socialising, with public house, general store (sometimes including a sub-post office), even primary school, adding variety. This is likely to be yet more widespread, and even involving duplication, in a **large village**.

Settlement location

Settlement locations chosen at random are likely to be very rare. It may be assumed, therefore, that early settlers deliberately searched for sites with distinct natural advantages, in situations offering the greatest potential for further growth and development. Given that any location was likely to offer both advantages and disadvantages, certain compromises may well be evident, with the best combination of positive factors likely to dictate the final choice. However, what must always be remembered is that most rural settlements were founded in entirely different socio-economic conditions than those of today, and whilst some may have been abandoned, the vast majority persist through inertia.

Original location factors include considerations of water and wood supply, defence, potential fertility, shelter, aspect, and dryness, plus trading and service opportunities.

1. **Water supply** - guaranteed to be fresh and regular - was the prime requirement for early settlers. Heavy to carry yet essential for life, it was used for domestic, agricultural, and economic activities, such as cleaning wool for barter. The relevance of water holes in arid locations, springs and wells elsewhere is, therefore, of great significance.

2. **Wood supply** in quantities sufficient for use as a building, furniture, utensil, and tool making material, heating and cooking fuel, even as animal fodder was likewise essential - with ease of felling an important consideration given the necessity to cut by hand. Forest-edge timber was, consequently, most prized.

3. **Defence** and so safe refuge was a necessary settlement function, given the civil instability associated with earlier centuries. Consequently, hill tops (noting the Celtic forts referred to earlier), rocky outcrops, and peninsulas suggest readily defensible sites. However, these were often inaccessible and exposed by comparison to, and adoption of, the natural moat effect inside river loops, at confluences, and on islands. It comes as no surprise to find far more settlements of the latter group continuing to thrive today.

4. **Potential fertility** was essential since early settlers depended entirely upon agriculture for survival. For example, if high soil fertility was combined with varied land then mixed farming utilising river flood plain meadows for cattle, friable clay loams for arable and rough pastures for sheep was possible. Certainly it is no coincidence today that more fertile areas such as East Anglia support higher village densities than relatively infertile regions such as the North Yorkshire Moors.

5. **Shelter**, **aspect**, and **dryness** have clear agricultural connotations and practical quality of life implications. Shelter from bad weather may be afforded by locating in rain shadow areas, on the lee slopes of ridges, or in valleys, which in the northern hemisphere would ideally facilitate south-facing comforts. For example, the south-facing aspect of vineyards along many Swiss Alpine valleys is necessary given the need for sunlight and warmth. Indeed, they are repeatedly found precariously perched opposite monotonous coniferous woodlands, which are less susceptible to the cold. Likewise dryness has both agricultural and domestic virtues with **dry seeking** settlements following valley sides or river terraces, notably both above the flood plain. This principle is taken to its logical conclusion in artificially drained and reclaimed areas where mounds may be constructed to elevate settlements above flood danger, as illustrated by the *terpen* of the Dutch polders.

6. **Trading** and **service opportunities** necessitate advantages of accessibility in order to allow ease of communication. River fording points (as illustrated earlier) provide early examples with major routeways likely to attract many settlements. Finally, the many villages established during the Middle Ages to serve other communities such as castles, monasteries, and cathedrals are worthy of particular note. Even today, research demonstrates lingering 'deferential traditionalism' in the **estate villages** of East Yorkshire (such as Sledmere, Birdsall, and Langton) despite the taxation realities of twentieth century England dictating increasingly fewer properties still occupied by 'tied' estate workers.

Village structure (morphology or form)

Although individually unique in site, situation, and characteristics, villages may be classified according to their structure, as determined by how they have grown into their existing shape and layout. Four factors which influence this are the **topography (relief)** of the site and situation, the

economics of the agricultural or trading system, the **culture** of the people as expressed in their religious and tribal groupings, and, not least, the **history** of the area dictating, for example, a need for defence.

Settlement compactness, for example, may vary according to site factors such as scarce land, or historical factors such as unstable politics. Clearly, therefore, one may find similar forms yet not necessarily the same reasons for them. Also one must always consider whether or not the structure observed today is in fact original or the outcome of progressive change and modification.

Fragmented (loose-knit) villages

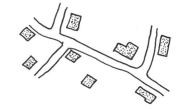

have no original nucleus due to the piecemeal clearance of woodland, without communal plans for development. Illustrated well by the *Drubbel*
villages of western Germany, they are unlikely to demonstrate much community spirit unless a church or public house added later engenders this.

Nucleated (clustered) villages

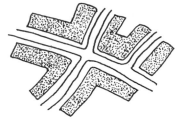

form usually on route centres.
Known as *Haufendorfs* (literally
'heaped up villages'), in central
Europe, their dense layout and
clear marginal definition may
reflect this trading function or, conversely, a need for defence. They are repeatedly associated with open field systems of strip farming (to be discussed later).

Linear (street) villages such as

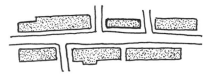

the *Strassendorfs* of the Black
Forest in Germany are elongated
along roads, rivers, ridges, or valleys.
Once again strip farming may be of relevance with the medieval strips cultivated as a 'back field' to each of the dwellings along the street.

Green (open space) villages are nucleated around a central open space (green) which may include a church or pond. The shape of the open space may determine their designations as, for example, the round *Rundlings* and cigar-shaped *Angerdorfs* of eastern Germany. In Britain they date back to

Saxon times when defensive origins seem probable. However, market and 'merriment' functions should not be discounted.

NB: African Masai and Zulu **kraals** are a contemporary equivalent with the central enclosure providing safe refuge for both inhabitants and their livestock.

Double villages are normally in two halves separated by a river, steep slope, or whatever. Qualifying adjectives may supplement the single name (such as High and Low Hutton in East Yorkshire) but this is not always the case. Nor can it be assumed that the two parts developed as a single community.

The distribution and pattern of rural settlement

Examining village structure demonstrates the study of settlements as separate entities. The alternative approach is to consider settlements as integral components of the environment; in other words, their spatial relationship to each other can be investigated – their locational characteristics analysed, explained, and even measured. We will study two elements of this – rural settlement distribution and pattern.

SELECTED DEFINITIONS

Distribution concerns how rural settlements of any size are spread across the countryside. Where are the settled areas? Where are the unsettled? Indeed, what are the limits of settlement?

Pattern concerns the character of the rural settlements themselves – sometimes irrespective of site. Settlements, for example, may be **dispersed** (as isolated dwellings) or **nucleated** (as villages).

Rural settlement distributions tend to vary from very sparse to very dense. Their spatial organisation may be clustered, random, or regular. Indeed, these terms apply when distribution is measured using **nearest neighbour analysis**.

Certainly, generalisations frequently prove to be valid. For example, the flatter the relief the more numerous the settlements. Likewise, fertile soils encourage settlement, just as inaccessible, hostile topography deters it. Indeed, variable relief, soil fertility, and water availability leads to clustered and/or random distributions. Settlements, therefore, congregate in

agriculturally favourable locations. Finally, uniform or flat relief encourages
regular settlement distributions.

NEAREST NEIGHBOUR ANALYSIS

This discussion of rural settlement distribution has been entirely subjective in that
there has been no measurement of, say, the degree of clustering or otherwise.
Quantitative analysis is, however, possible and a valid tool when accurate comparison
of contrasting areas is required. Nearest neighbour analysis (referred to in *Population*)
allows one to measure rather than just describe distributions.

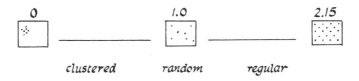

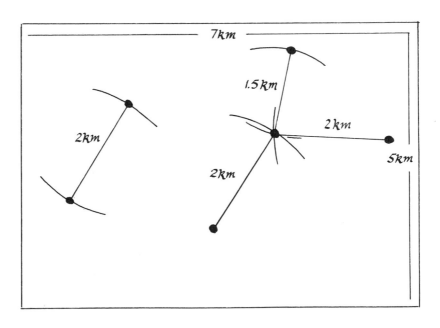

The nearest neighbour index is calculated by the formula:

$$\mathbf{Rn} = \mathbf{2d}\sqrt{\frac{\mathbf{n}}{\mathbf{A}}}$$ where

$\mathbf{Rn}$ = the nearest neighbour index

$\mathbf{A}$ = the size of area concerned

$\mathbf{d}$ = the mean distance between settlements (taken as an average of the distance between them)

$\mathbf{n}$ = the number of settlements

Values for $\mathbf{Rn}$ range from 0 when there is no distribution at all to 2.15 where settlements are uniformly spaced. An $\mathbf{Rn}$ value of 1.0 represents a purely random distribution. Consequently, values less than 1.0 represent clustering - the smaller the value, the greater the degree. Values more than 1.0 are described as regular - the higher the value, the greater the degree of regularity.

$$d = 2 + 1.5 + 2 + 2 = \frac{7.5}{4} = 1.9 \text{ km}$$

$$\mathbf{Rn} = \mathbf{2d}\sqrt{\frac{\mathbf{n}}{\mathbf{A}}}$$ where
$\mathbf{A} = 5 \times 7 = 35 \text{ km}^2$
$\mathbf{d} = 1.9 \text{ km}$
$\mathbf{n} = 6$

$$= 2 \times 1.9\sqrt{\frac{6}{35}}$$

$$= 3.8\sqrt{0.17}$$

$$= 3.8 \times 0.4$$

$$= 1.52 \quad \text{hence a regular distribution.}$$

Criticisms of nearest neighbour analysis

1. It only deals with selected settlement sizes such as villages.

2. The determination of the centre of the settlement is subjective. Even when prestated, such as the village church, there is still room for error.

3. No account is taken of the physical or human landscape. For example, straight-line distances are measured regardless of roads, physical barriers, and so on.

4. It gives no indication of pattern whatsoever.

Rural settlement patterns by contrast do not invite such logical generalisations because they result from numerous and frequently complex social, political, and historical factors which are intrinsically 'woolly' and difficult to measure. Patterns are also liable to change as agricultural practices evolve, and populations develop and grow.

Further difficulties of analysis arise from the fact that patterns are frequently blurred. Only occasionally will clear dispersal or nucleation occur. Indeed, intermediate patterns with elements of both are, alas, the norm.

Many theories attempt to account for different rural settlement patterns. Some focus on agricultural systems, others on the need for defence. Population pressure and culture have also been proffered as explanatory. However, it is likely that all have influence to a greater or lesser extent depending upon the circumstances.

Agriculture is certainly of great relevance given the repeated evidence of Saxon nucleation whereby a communal agricultural system of strips within open fields surrounding a village could be contrasted vividly with earlier Celtic practices of piecemeal individual cultivation dictating dispersal. There is also a link between nucleated villages with arable cultivation on the one hand and dispersed isolated dwellings with pastoral farming on the other. Remember - livestock requires far more daily attention than crops, so necessitating ready access for the farmer. Similarly, intensive agricultural practices encourage nucleation whilst extensive farming allows dispersal. Certainly, agricultural change as discussed later emphasises the relevance of farming to explaining rural settlement patterns.

Protection afforded by nucleation (in villages) may reflect social and political instability. The rarity of evidence for isolated dwellings in England during the Dark Ages, for example, would seem to confirm this logical assumption.

Population pressure, likewise, has logical consequences - with nucleation resulting from lack of space. In contemporary China, for example, some villages are so compact that up to half their inhabitants may live under the same roof!

Culture may lead to different patterns in similar environments. Strong tribal, clannish, or even family ties, for example, may encourage nucleation with dispersal more likely where cultural bonds are weak.

The factors above clearly have relevance in helping to clarify rural settlement patterns. However, they don't explain how or why such patterns might change over time. For example, does dispersal result from the break-up of nucleation or does it pre-date it? Certainly, in Europe the former seems likely, because as civilisation progressed the need for protection and cooperation as afforded by village life declined. Indeed, this dispersal was encouraged further by the disintegration of feudalism.

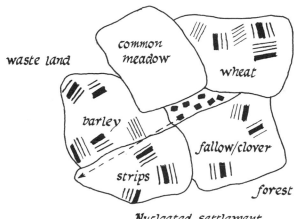

Nucleated settlement
(pre-enclosure)

///|/// strips cultivated by one man   ■ buildings

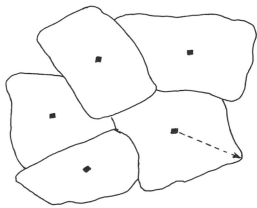

Dispersed settlement
(post-enclosure)

◄ - - - - - distance to furthest point of farmland

Just as the French Revolution (1789) saw the lands of large estates divided amongst the *proletariat*, settlement dispersal was also seen in England, due to the late eighteenth and early nineteenth century Parliamentary Acts of Enclosure, encouraging more efficient farm-based agriculture.

Elsewhere in the world similar temporal changes are of note. For example, whilst the pros and cons of the British Empire can always be guaranteed to stimulate debate, few can deny that colonial stability allowed safe settlement dispersal. This can be contrasted to contemporary Israel where nucleation in, for example, *kibbutzim* serves both agricultural and defensive functions.

In conclusion, therefore, when dealing with rural settlement patterns we must always acknowledge that there may be a historical factor to consider. The contemporary pattern may well prove to be the outcome of a long period of change.

CENTRAL PLACE THEORY
(*Central Places in Southern Germany*, W. Christaller, 1933)

Central place theory aims to explain the spatial organisation of settlements and their hinterlands, in particular their relative locations and size. Christaller's theory was based on the assumption that there was some sort of order in the pattern and functions of settlements. (He subsequently tested his theory in southern Germany.) Settlements were classified according to their functions, and the relationship between them and their hinterlands examined. Christaller was suggesting that there was an overall organisation to the system of settlements, and hinterlands, and was concerned with their relative rather than absolute position. He studied the whole range from what we would describe as hamlets to cities, regarding them as service centres to the surrounding areas (hinterlands/market areas/spheres of influence/urban fields).

Christaller's theory necessitated various, now familiar, assumptions in order to simplify reality. Landscape (topography) was simplified to an unbounded, featureless (isotropic) plain with uniform soils, climate, and resources and an evenly distributed population with equal ease and opportunity of movement in all directions. Transport was not, therefore, confined to networks, and transport costs would only vary with distance - so demonstrating the friction of distance. Just as with transport costs, it was assumed that demand would decrease with increasing distance from the central place and that there would be a maximum distance which a consumer would be prepared to travel in order to obtain a good or service - the **range of the good** marking the outer limit of the market area. Christaller also appreciated that any goods or services would have a minimum demand (or market size) below which overheads could not be met. Obviously, any supplier would attempt to obtain a far larger market than this so-called **threshold** in order to maximise profits, thus necessitating locating as far as possible from other suppliers. The simplified motives of the supplier were matched by consumers who were assumed to have identical needs and tastes and seek to obtain these from the nearest supplier.

Consequently, hexagonal market areas achieved these objectives - efficiently covering the plain with no gaps or overlaps. Furthermore, different goods and services would have market areas of different sizes. **Low order** goods such as groceries commanded small market areas - **high order** goods such as furniture, relatively large. The larger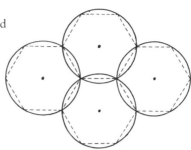
the settlement, the more goods and services provided - hence low order centres with small market areas and higher order centres attracting custom from farther afield. Consequently, a hierarchy of settlements and a mesh of hexagonal market areas emerges, with settlements of equal size being evenly spread, offering identical functions and servicing identically sized hinterlands. Clearly there would be many more smaller settlements providing **convenience** goods and services to small hinterlands. There would then be progressively fewer and fewer settlements of increasing size commanding larger market areas for infrequently required, more specialised **comparison** goods and services. These larger settlements would, however, provide all the goods and services of lower order centres - thus ensuring that large centres serve large market areas with a great number and variety of services, whilst smaller centres command smaller hinterlands with few and limited services.

The spatial arrangement of these **central places** are described by **k** values indicating the number of centres dominated by another centre and the relationship between the number of hinterlands of each order.

k = 3 (the marketing principle)

This principle most efficiently served the needs of consumers. Each settlement is surrounded by six smaller settlements of the next order down, situated at the hexagon corners. Each high order settlement, therefore, serves the equivalent of three lower order ones - hence k = 3.

x serves 6 smaller centres, each of which has a choice of 3 larger centres. Consequently, on average, one-third of each lower order settlement uses each higher order one. Therefore, x serves one-third of six smaller centres, plus its own smaller lower order hinterland.

RURAL SETTLEMENT

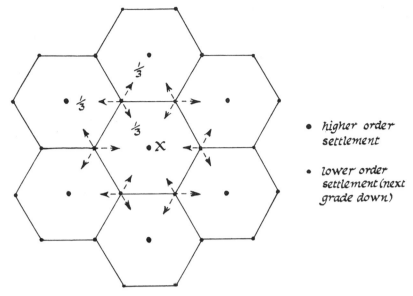

- • higher order settlement
- • lower order settlement (next grade down)

Hence k = 1/3 + 1/3 + 1/3 + 1/3 + 1/3 + 1/3 + 1 = 3

k = 4 (the traffic principle)

This principle maximised the number of settlements along straight lines, thereby facilitating movement. The lower order centres had to be on the hexagon flats.

Each lower order settlement is equidistant from only two higher order centres. Consequently, each larger centre serves half the population of six smaller centres, plus its own small lower order hinterland.

Hence k = 1/2 + 1/2 + 1/2 + 1/2 + 1/2 + 1/2 + 1 = 4

k = 7 (the administrative principle)

This principle eliminates any shared allegiances whereby a small settlement was administered by more than one larger centre.

The larger and again re-orientated hexagonal hinterland of the higher order centre encloses the six lower order ones. Consequently each larger centre serves six smaller ones, plus its own small lower order hinterland.

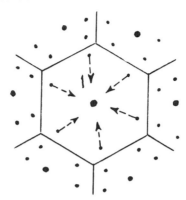

Hence k = 1 + 1 + 1 + 1 + 1 + 1 + 1 = 7

Testing Christaller's Theory

On testing the theory in southern Germany, Christaller assumed that the marketing principle (k = 3) was the main determinant of a system of central places and indeed finally concluded that this was so, with k = 4 and k = 7 systems as *'secondary laws causing deviation'*. Christaller did little testing outside of Germany, but numerous other studies have examined the theory and specific components of it. It has been demonstrated, for example, that threshold populations for different goods and services support Christaller's logic, although, inevitably, exceptions such as the higher order good in the village shop are reported. Likewise, studies of range suggest the obvious – that people travel shorter distances for lower order goods, but larger distances for higher. Again, however, exceptions are always found, increasingly so nowadays with out-of-town shopping complexes. Proving Christaller's suggestions regarding distinct hierarchies of settlements, however, has always proved more difficult, with variable results.

Appraising central place theory

Criticisms of the theory (and modifications of it such as by A. Lösch) centre on the truism that the real world is far more complex than the initial assumptions allow and consequently the proposed settlement patterns are palpably unrealistic. Homogeneous regions just do not exist. Transport networks and physical geography dictate movement, not equally easy in all directions, but channelled along networks where choice of mode is an important consideration, especially when one considers the convenience afforded by the private car. Furthermore, humans are not totally rational. In the real world producers may not always aim solely to maximise profits for perhaps social or philanthropic reasons. Likewise, customers may not always shop at their nearest store. Indeed, retailing habits and patterns are changing markedly with, for example, multiple (chain) stores such as *Sainsbury* or *Marks and Spencer* severely distorting market areas. These organisations can generate a demand and even support initially loss-making outlets until the threshold market is eventually realised. Most of these chain stores command disproportionately large hinterlands luring increasingly mobile customers by low prices (often using **loss leaders** - big brand necessities sold at sometimes below cost in order to attract business), great variety, standardised quality, credit facilities, and not least the convenience of 'everything under one roof'. Indeed, Britain now has hundreds of out-of-town **retail parks**, such as Willerby Park outside Hull, including a 'super-breed' of American style **regional shopping centres** based upon the principle of shopping as a 'leisure concept'. These complexes, such as Meadowhall near Sheffield and Thurrock Lakeside on the M25, command major regional hinterlands. However, whether chain stores, local, or regional shopping centres, the resultant by-passing and eventual bankrupting of many low order service outlets must invalidate, nowadays, much of central place theory. The working and domestic circumstances of most families have simply changed too much since the 1930s, with daily shopping for fresh perishables no longer necessary due to fridges, processed foods, and so on. Likewise, the realities of topography, and not least the fact that similarly sized settlements do not necessarily share the same functions, conspire further to prohibit Christaller's regular hexagonal market areas.

However, despite all these valid criticisms, it would be crass to dismiss central place theory out of hand. Just as he noted settlement regularity in southern Germany, so have others, successfully, applied these principles to other regions of flat relief, such as East Anglia. The value of Christaller is that deviation in the real world can be measured against his idealised pattern. Therefore, we can compare reality to the theoretical and then seek to explain why it is different, so promoting a real understanding of the geography of settlement and especially retailing. Christaller's research has identified the element of order in the integrated systems of central places and market areas, dispelling forever the previous misconceptions that the settlement and its hinterland could be viewed in isolation from other settlements and their hinterlands.

17

URBAN SETTLEMENT

The definition of **urban** is not straightforward because urban areas differ widely in size, appearance, function, and structure. Indeed, when does a rural area become urban? Are urban outskirts definable as such? Certainly, definition by population (as suggested by the UN) is fraught with difficulties, especially given numerous national adaptations to the system. Suffice it to say that the distinction between, say, a village and a town is imprecise and, as before, descriptive criteria with reference to popular local perceptions may well clarify best what we determine as rural and all that we call urban.

Towns tend to be distinguished by a wide range of non-agricultural services and functions. Chain stores may be present and certainly shops specialising in the same goods in competition (which may be likened to competing churches of the same denomination). Education provision, from primary to tertiary, may be present, along with specialised recreational services, such as leisure centres and cinemas. Finally economic specialisation, particularly industrial, may be a characteristic, such as in the association of Scunthorpe with the manufacture of steel. **Large towns** are likely to demonstrate an increase both in size and variety of services and functions, yet weakening social bonds, regardless of the close proximity of diverse groups of people. Yet when do large towns become cities? *'A town with a cathedral'* is questionable too often, but in Britain a Royal Charter of Incorporation, such as granted to Sunderland in 1992, clarifies **city** status administratively. Certainly *'considerable diversity of function'* is the ideal descriptive theme. They are much larger than towns, with varied economic functions, regional administrative offices, transport termini, and major financial institutions. Unfortunately, the *'anonymity of the city'* may well be an unhappy social reality. Larger urban entities than cities command varying definitions according to the particular country. In Britain, for example, we recognise the amalgamation of merging towns and cities as **conurbations** with the principal commercial centre acknowledged as the focus - hence Leeds as the centre of the West Yorkshire conurbation. The USA refers to **metropolitan** areas in this respect, but now has to acknowledge even greater sprawls - super-metropolitan regions such as the famous **megalopolis** of *Bosnywash (Boswash)*, a 1000 km urban belt from Boston, through New York, to Washington.

The relationship between town and country

An important aspect of central place theory was the recognition that towns and larger urban settlements were not isolated entities but linked inexorably to the surrounding area. These physical and socio-economic links involve interactions between the urban and surrounding rural area and *vice versa*. The interactions may be of people, goods, money, even information. Indeed the urban centre may provide the surrounding area with various goods and services - this tributary area being referred to, often, as its **zone** or **sphere of influence**. However, **urban field** is now the most generally accepted term and it covers the entire area, both urban and rural, influenced by the central town.

The links (functional relationships) within urban fields

As just indicated, the interactions between a town and its surrounding area are likely to be two-way. For example, food products for the town, especially perishable horticultural and dairy items, may be provided by surrounding farmers. Conversely, the town may serve as a collection and distribution (market) centre for much of their agricultural produce. Similarly urban industries may depend upon raw materials from the surrounding area in order to manufacture finished products, some of which will be sold within this urban field. Certainly, the town is likely to be an influential provider of services to the surrounding area - be they retail (especially of higher order comparison goods), professional (such as lawyers), health care (hospitals), educational (notably secondary schools and colleges), or entertainment. Clearly, the two-way process should always be considered. Just as the town's clubs, theatres, cinemas, and restaurants attract customers from outside the town, the local countryside, conversely, is likely to offer recreational opportunities for the urban population. Perhaps the most readily apparent interaction, however, is the regular circulation of commuters (see *Migration*). These urban workers travel daily from, and return to, the environmental advantages of rural dormitory settlements.

The size and shape of urban fields

Unlike the predictable hexagonal market areas of central place theory, real urban fields are likely to be irregular both in size and shape. For example, an urban field might be truncated by the coast or elongated along a valley. Certainly, towns demonstrating strong functional linkages with the surrounding area are likely to have larger urban fields than those with weaker links. However, numerous factors influence their nature – the town's functions, services, and employment opportunities, not least its 'character', will all be of relevance. Indeed, simplistic assumptions relating, in positive correlation, the size of population to the area of urban field must be avoided. P.R. Odell's (1957) comparison of Melton Mowbray with nearby Coalville is repeatedly quoted in this context. The smaller, historic market town of Melton Mowbray commanded a far bigger urban field than the larger, but less accessible, relatively degraded mining centre of Coalville.

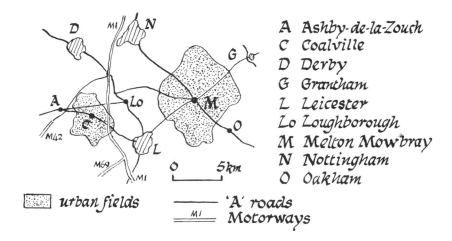

A Ashby-de-la-Zouch
C Coalville
D Derby
G Grantham
L Leicester
Lo Loughborough
M Melton Mowbray
N Nottingham
O Oakham

▨ urban fields

——— 'A' roads
══MI══ Motorways

A town's economic structure is linked to the urban field in that it will serve the surrounding area – not just its own population. Consequently, a proportion of the town's labour force is concerned with producing goods and services for export (**basic** or **city-forming** workers), so earning money from the surrounding area to buy new materials, food, and so on. The remaining (**non-basic** or **city-serving**) workers serve the urban population itself. The ratio between basic and non-basic workers is thought to determine the size of the urban field – a high proportion of basic workers thus indicating a large urban field, such as in a tourist resort.

However, one must remember that the size and shape of urban fields will change over time. They are dynamic, responding to changing technological, social, and economic circumstances. For example, improved transport links such as a motorway or dual-carriageway will widen urban fields. Conversely, a railway closure or reduced bus service will shrink them. Even seemingly trivial, ephemeral social factors, such as the fickle popularity of 'night spots', will influence urban fields.

The delimitation of urban fields

Each town will have a succession of urban fields, of differing shapes and sizes, representing different services and functions. Indeed, just like Christaller's model, there will be a hierarchy of them with the urban fields of the largest settlements encompassing those of the smaller ones.

If one can delimit urban fields then our understanding of the interactions between settlements would be improved – consequently aiding planning and local authority administration. So how is this done?

Fieldwork, whether urban or countryside orientated, will involve collecting information on a variety of characteristics or representative indices. The more indices used, the greater the accuracy. However, logistical difficulties normally prohibit a varied survey, necessitating the use of just one index to express the town's links with the surrounding urban field. A local newspaper can be this ideal source, particularly if the origin of 'small ads' is studied, because births, marriages, deaths, lost kittens, and so on rarely command anything other than local interest.

Clearly there may be a necessity for more accurate delimitation of urban fields, particularly if fieldwork research on the increasingly less likely retail delivery areas, public transport services, and, say, church, club, and clinic catchments reveals vagaries or contradictions. Both the **gravity model** (see *Migration*) and **breaking point theory** (despite the dubious assumption that population, solely, determines influence) are quantitative measures adopted.

Breaking point theory is particularly useful to planners and administrators who need to establish a clear demarcation line between two settlements. For example, when two urban fields overlap and towns compete for the same people, or when two urban fields fail to meet, creating an unserved vacuum.

BREAKING POINT THEORY (W.J. REILLY'S LAW OF RETAIL GRAVITATION, 1933)

The use of the **gravity model** to estimate the extent of an urban field is now generally accepted. Just as the interaction between two towns decreases in proportion to the square of their distances apart, so the influence of a town over its surroundings is thought to decrease in proportion to the square of the distance from the town.

However, Reilly's **law of retail gravitation** allows us to measure the limits of trading areas and so the extent of a town's influence on its surrounding area. For example, equally sized towns would exert equal influence on the surrounding area – hence the **break point** halfway between the two. Normally, however, town sizes vary – necessitating use of Reilly's formula:

$$\text{Distance of break point from a (where town b is smaller than town a)} = \frac{\text{distance a to b}}{1+\sqrt{\dfrac{\text{population b}}{\text{population a}}}}$$

NB: Always quote the break point from the largest town.

Urban settlement hierarchies

Many geographers believe that settlements have grown in a logical order in that their sizes, and even functions, relate to an overall regularity. Random settlement development is, therefore, discounted.

G.K. ZIPF'S RANK SIZE RULE (1949)

Zipf expressed this suggested relationship in precise mathematical terms. His **rank size rule** stated that *'if all the urban settlements in an area are ranked in descending order of population, the population of the nth town will be $1/nth$ that of the largest.'*

Therefore $P_n = \dfrac{P_1}{n}$ where

P	=	the population
n	=	the rank
P₁	=	the population of the largest city

where P = the population, n = the rank, P_1 = the population of the largest city

For example: P_1 = 9 000 000

$$P_2 = \frac{9\ 000\ 000}{2} = 4\ 500\ 000$$

$$P_3 = \frac{9\ 000\ 000}{3} = 3\ 000\ 000 \text{ and so on}$$

Plotted on normal graph paper, this gives a geometrically curving line. Plotting on logarithmic paper, however, straightens it, so allowing easy visual comparisons of reality against the mathematical projection.

The settlements of the USA at this time conformed to the theory. However, the largest settlements were, predictably, less numerous than smaller ones and, of course, there were many centres of similar size. Consequently, Zipf's rank size rule is normally modified to a **stepped order** in which a number of settlements are found at each level.

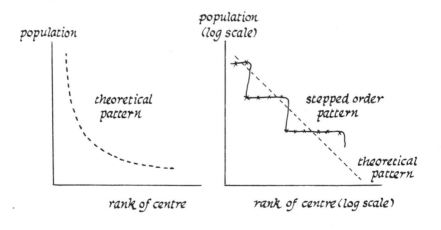

Other modifications include the **binary pattern** whereby a number of similarly sized centres dominate the upper end of the hierarchy with normal rank size progression thereafter. The **primary pattern** is more usual, however, with, again, settlements low in the hierarchy conforming to the theoretical rank size progression, but the largest city being up to six times bigger than the centre ranked second. M. Jefferson's *Law of the Primate City* (1939) suggested that this was only natural – a primate city, such as a country's capital, might be expected to outgrow all others as a self-perpetuating growth pole attracting investment, talent, and ever increasing services and functions.

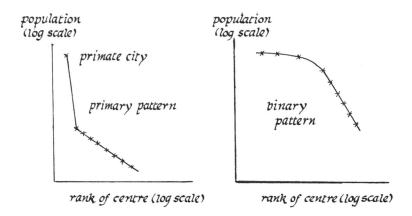

Appraising the rank size rule

Evaluation of the rank size rule is based upon empirical studies of numerous countries. Three distinctive groupings emerge:

1. Countries which displayed very close relationships to the theoretical log normality included Sweden, Switzerland, and the USA (at the time of Zipf up to the early 1970s, after which a binary pattern emerged). What was of particular note was that these countries were normally larger than average with a long history of urban growth. They were also economically complex. This latter point is illustrated well in countries with a federal government structure, such as Switzerland. This prevents the dominance and centralisation of economic and political power in one large urban unit.

2. Countries which displayed a binary pattern had two or three cities disproportionately larger than the rest and notably complementing rather than competing with each other. For example, Brazil's former capital, Rio de Janeiro's commercial, tourist, entertainment, and lingering administrative functions complementing Sao Paulo's role as an expanding industrial city.

3. Countries which displayed a primary pattern were normally smaller than average with a short history of urban growth. ELDCs were repeatedly of note, often with the main city acting as the servicing, collection, and distribution centre for the 'mother' country (former colonial power) rather than servicing to the full their respective regional economies. Urban primacy was also tending to occur in countries with markedly uneven population distribution, again reinforcing the theme of a dominant city (usually the capital) and its region enjoying disproportionate economic and political power at the expense of the peripheral regions. Santiago, Chile's capital, in the country's central region, illustrates this point particularly effectively.

A concluding note should emphasise that the rank size rule is based upon the analysis of actual (empirical) data and is not a body of theory. Practical difficulties restrict its study. For example, unless you use population statistics from, say, an atlas (therefore dated), how do you determine city population, given that suburban development may cross administrative boundaries? Of most note, however, is that despite plausible explanations of exceptions to the rule, such as the primary and binary patterns, attempts to explain adherence to the theoretical have been largely unsatisfactory.

Urbanisation

Urbanisation occurs when an increasing proportion of a country's population lives in towns and cites rather than in the countryside. It represents transfers from rural to urban lifestyles.

The origins of urban living have been (radio-carbon) dated to around 6000 BC in the riverine areas of South-West Asia. Such early urban lifestyles were possible due to increased agricultural production allowing a surplus for trade, together with more sophisticated social organisation and division of labour. Hence the first pre-industrial cities acted as central places for collection, storage, and distribution. Additional to this marketing function, however, was their role as religious, social, and political foci for the surrounding population. For example, the Harappan civilisation of the Indus Valley (2300-1750 BC) had 70 urban sites covering about 1.3 million square kilometres, the best known being *Mohenjo-Daro*. These centres were large and well planned. Indeed, uniformity of town plans, architecture, culture, and script is clearly evident.

However, urbanisation is a relatively recent phenomenon. Even in medieval Europe the mass of the population were still rural dwellers. By 1800 only 3 per cent of the world's population lived in settlements over 5000 people. Since then, however, the proportion has increased markedly – with a global urban majority now likely next century.

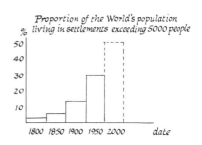

Proportion of the World's population living in settlements exceeding 5000 people

The reasons for urbanisation are both numerous and complex. Agricultural change is particularly relevant and best illustrated by the onset of the Agricultural Revolution in eighteenth and nineteenth century Europe, whereby mechanisation and other innovations reduced demand for rural labour, yet produced food surpluses for the growing urban populations. The concurrent Industrial and Transport Revolutions accelerated the migration of rural dwellers into the urban areas, where industrial agglomeration in locations well endowed with mineral and energy resources allowed economies of scale and shared infrastructure. Local, regional, and national communications progressed through railways superseding waterways, whilst more efficient global shipping allowed the *New World* to be harnessed for further food production. A self-perpetuating theme becomes apparent – increasing urban populations represent a focused market demanding not just industrial products but services too. Certainly increased trade, transport opportunities, and rising living standards allow more personal aspirations to be met. Educational, social, and cultural opportunities are likely to be more varied and stimulating in urban areas – indeed the *'bright lights syndrome'* persists to this day, and is of probable significance when trying to explain a process associated with notable advantages, but many disadvantages too.

However, there are dangers in such excessive cultural and temporal generalisation. For example, the rate and nature of contemporary urbanisation varies throughout the globe. Indeed, affluent commuting-inspired **counter-urbanisation** (as depicted in the inappropriately named **cycle of urbanisation**) is evident today in many EMDCs.

CYCLE OF URBANISATION

1. Rapid urbanisation due to Agricultural and Industrial Revolutions.

2. Reaching maximum urban proportion of about 80 per cent.

3. Counter-urbanisation (decentralisation).

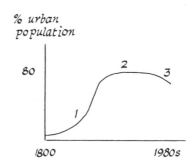

NB: This Eurocentric model is based upon European experience and so cannot apply to every country.

ELDCs by contrast face urbanisation rates far in excess of western Europe during the Industrial Revolution. Social and cultural circumstances differ too. Family size, for example, in South-East Asia has increased with rising incomes, rather than decreasing as it did in western Europe. ELDCs frequently see not only large rural to urban migrations, but rising natural increase too as mortality declines with improving hygiene, health care, and so on. Certainly, a vast potential for accelerating urbanisation exists in numerous ELDCs. India, for example, has over 50 million landless peasants! Statistics focusing on urbanisation in the South, even with carefully conservative estimates, stagger and appal. Just as the majority of current **millionaire cities** are in the South, so too will be most projected **super cities** next century. Focusing on just one of the three anticipated in India, Calcutta already has over half of its people squatting illegally in sprawling **shanty towns** (known locally as *bustees*). A further 200 000 are thought to sleep on the pavements. When further degradations such as the 'hot bed' tenement 'hotels' and ubiquitous pit latrines are contemplated, one wonders why the urbanisation process continues unabated. Perhaps general observations might best emphasise the reasons.

Whether a permanent or temporary migration, the risks of the city may be far more preferable to guaranteed poverty in the countryside. Only a fraction of migrants can expect to find industrial work, for urbanisation is occurring ahead of the industrialisation needed to absorb them. Yet the informal sector (black economy) provides infinite opportunities. Whether the desire is money, ambition, freedom, education, or health care, the lure of the city is strong. Urbanisation is, consequently, an understandable process reinforced by repeated illustrations of ELDC cities as relatively safe havens from natural and man-made disasters such as drought and civil war respectively.

In conclusion, therefore, we can clearly relate economic and technological progress to wealth creation, with consequent potential for support of the arts and other cultural activities. This emphasises much that is positive about urbanisation. However, the process is more often associated with problems. Indeed, repeated parallels may be drawn between North and South, providing appropriate adjustments are made to account for specific realities, not least acknowledging that urbanisation is generally concluding in the former whilst ongoing in the latter. Water supply, waste disposal, and general infrastructure provision, noise, air and water pollution, traffic congestion, and structural decay, not least poverty, are all familiar global urban realities. It is only emphasis and specifics which differ. For example, **urban renewal** in Leeds will contrast markedly to **site and service**

schemes adjacent to Calcuttan bustees - yet both share the admirable planning ideal of improving the lot of the urban dweller.

Urban structure

This is the physical arrangement, layout, or morphology of an urban area. Its study involves the description, analysis, and explanation of land-use patterns, urban district functions, building density variations, and so on.

Urban land-use tends to be a complex mosaic collectively determined by years and years of evolution. Numerous individual and corporate decisions by residents, planners, industrialists, councillors, and so on will, therefore, dictate the layout.

Arguably, it is economic factors which most influence urban land-use decisions. In theory, any activity, whether retail or commercial, industrial or residential can derive **utility** from any site - a utility measured by the rent (reflecting land value) that an activity is willing to pay. Land values arguably reflect accessibility - the centre of the urban area as the focus of transport routes being the most accessible point with the highest land value. Retail and commercial functions demanding maximum customer potential will be willing, therefore, to pay highly for central locations - thus excluding other land-uses. Decreasing accessibility from the centre, with corresponding declining land values, allows an ordering of land-uses related to rent affordability. This is expressed in the theory of **bid-rent analysis**.

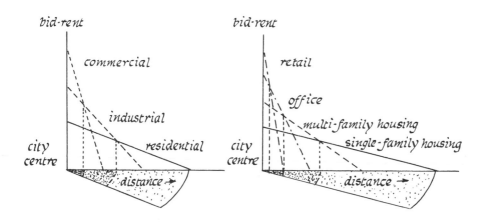

In reality, a uniform decrease in land values from the centre is a nonsense because of radial routeways and so on. A more realistic pattern is, therefore, one of an urban land value surface with peaks and troughs reflecting variations in accessibility. Peaks of high land values would correspond to major route intersections, such as radial routes from the centre joining an inner ring road.

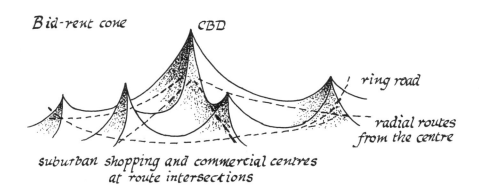

Bid-rent cone   CBD

ring road

radial routes
from the centre

suburban shopping and commercial centres
at route intersections

The areas of lowest land values will be those distant from the main routeways or lying between them. Also, it is notable that a strong correlation exists between bid-rent curves and building heights. High central land values necessitate tall buildings, making the best use of limited space. As land values diminish, so do building heights and densities. A less intensive use of land is, therefore, possible.

However, accessibility means different things to different activities. For example, an industry producing for a national market might view a peripheral suburban location adjacent to a motorway junction as 'most accessible'. The assumption that accessibility to city centre shopping and work as being paramount, compared to, say, nearness to scenic areas away from industrial land-use has to be questioned, especially given the personal mobility (or immobility in city centres!) of the car and out-of-town shopping centres. Not least is the paradox of the lowest income groups often located in inner city areas on high value land. This is usually explained in terms of transport costs to work in the central area being critical – the houses having been built in an era of low land values. Higher income groups with more money for travel, therefore, occupy low density housing on large areas of lower value, suburban land.

Functional areas within the city

Any large city fulfils many functions and provides numerous services. Inevitably, it is a centre of commerce, industry, administration, retailing, transport, education, and entertainment, not least the home for thousands of people. However, despite a great deal of mixing, different urban functions tend to arrange themselves in a series of distinctive land-use zones or functional areas such the central business district (CBD), industrial, and residential districts.

The central business district (CBD)

Subject to more research than any other functional area, the CBD to outsiders is often synonymous with the city itself. It is likely to be located at or close to the geographical centre of the total urban area, yet occupy a relatively small proportion – as little as 5 per cent. It may well include the historic core, stimulating planning conflicts given the impracticality of, for example, medieval street patterns in the age of motor vehicles. Certainly the qualifying adjective 'high' is repeatedly appropriate – high building densities and height, high land values, high traffic and pedestrian densities, high class retailing. Department and chain stores, legal and financial services, public administration and offices, not least varied entertainment facilities, all characterise the CBD. Indeed, it is not just a distinctive land-use area in itself but likely to demonstrate internal functional zoning too (see later).

Delimiting the essential characteristics of the CBD (accessibility, building heights, traffic and pedestrian flows, location of residential population, if any, land and property values, even people's perception of it) involves measuring and mapping. Yet it is not easy to use these elements in practice in order to delimit the CBD's extent, because they are concentrated at the core and gradually diminish towards a rather 'fuzzy' or indefinite boundary. The edge of the CBD often merges into an area of blight, where derelict and obsolete property separates it from inner city residential areas. Sometimes referred to as the twilight zone, but more normally the **zone of transition**, it is characterised by redevelopment and renovation.

Indeed, some writers distinguish two CBD zones in terms of land-use intensity – the inner **core** of high intensity, with strong functional linkages between functions, surrounded by a **frame** of less intensive land-use, with functions having little in common with each other.

R.E. Murphy and J.E. Vance (1954) devised the most objective and widely applicable method of delimitation based on land value decline outward from the core. A study of nine American cities showed that a land value of 5 per cent of the peak land value, corresponded closely with the extent of the CBD determined by measuring and mapping the characteristics described earlier. Using this method in non-American cities has shown a less steep decline in land values from the so-called **peak land value intersection** (PLVI) and higher percentage values, but the method is clearly of some value. Practical difficulties, however, in persuading local authorities to release valuation data, frequently necessitate a compromise study combining available land valuation information with land-use mapping of typical CBD characteristics.

However you delimit it, the CBD must not be regarded as static. Murphy and Vance, for example, noted that despite the high investments in buildings and land, the boundaries of the CBD will gradually shift as economic conditions change. They referred to a **zone of assimilation**, where extensive redevelopment spread shops, offices, and so on into former residential zones. Conversely, a **zone of discard** was characterised by closures, dubious services, warehouses, and wholesalers – an area, therefore, with much vacant property and a run-down appearance. It is also notable that the PLVI will drift in response to these changes – away from the zone of discard in the same direction as the zone of assimilation.

Land-use within the CBD demonstrates its own **internal structure** – an order reflecting land values. The PLVI, for example, sees companies (usually retail) generating high profits from large turnovers. Hence the department and chain stores thriving on high pedestrian densities with corner sites being particularly valuable. Other high order comparison shops, such as women's clothing and shoe shops, cluster near the major stores to take advantage of the heavy pedestrian flows. Everyone benefits – the shopper has convenience and variety, whilst all retailers benefit from their proportion of the comparison trade.

Murphy and Vance devised another system of analysis in order to determine land-use changes away from this central area. They plotted four concentric zones, each 100 yards in width, around the PLVI. Within each zone, the percentage of land-use devoted to specific functions, at both street and upper floor levels, was calculated. The hypothesis of ordered adjustment to land values and distance from the PLVI proved valid, albeit not in neat concentric zones. Mutually dependent shops and services clustered, with legal, insurance, and financial services often associated with a clearly distinguishable district, as illustrated in the clear functional zones of Leeds' CBD:

- **Retail**
- **Legal/financial/insurance**
- **Medical/hospital**
- **Higher education - Leeds and Leeds Metropolitan universities**
- **Public administration - council**
- **Industrial/wholesale**

Usually offices do not require high levels of pedestrian visibility, so can be 'tucked away' behind shops where land values are lower or in high rise blocks behind the main shopping streets. Most likely to break the neat theoretical zones will be the inevitable **ribbon developments** of shops following routeways to prominent car parks or bus and railway stations. A progressive downgrading of functions continues to the margins of the CBD, characterised by smaller traders with modest profits and the discount warehouses demanding more space for floor area and car parking.

Closer study of the zones reveals specialist shops, such as antique dealers, bookshops, and gunsmiths, often found in side streets near the main shopping area. This is because their trade is not dependent on impulse buying. Conversely, banks, building societies, and estate agents do need a high profile, hence their ground level location in the main shopping streets. Other offices and personal services often locate above retail outlets.

A final note should again refer to the dynamism of the CBD. Change is ever-apparent - pedestrianisation of shopping streets, multi-storey car parking, and so on. High land values, pollution, and congestion, along with improved telecommunications, are resulting in the American phenomenon of many retailing and office functions joining industry in peripheral, suburban locations. The **dead heart** of the city centre is increasingly apparent and demanding the attention of politicians and planners alike. Indeed, do the 'Meadowhalls', 'Thurrock Lakesides', even their smaller, local predecessors, threaten the very *raison d'être* of much of the CBD?

Industry in the city

The importance of a city's industrial function will vary considerably according to its regional economic setting. Dominant in some, it may be scarce in others, hence only broad generalisations may be made about the location and structure of urban industrial districts.

In Britain, for example, industry used to be far more common in the centre of cities. Indeed, its replacement by shops has left much of the remaining inner city industry as a relic feature occupying low grade sites within the zone of transition. However, a central location may still be advantageous, for skilled workers can be readily drawn from the whole urban area, as illustrated by the jewellery manufacturers of London's Hatton Garden. Also, CBD retailers may market centrally produced goods such as fashion textiles. Certainly, a central location can aid rapid distribution of perishables, such as local newspapers and bakery products. The decline of so much central industry relates to the contemporary difficulties of this environment, such as illustrated by decaying buildings, site constrictions, and traffic congestion. Frequently, urban renewal gives central manufacturing an opportunity to relocate in purpose-built industrial estates, usually in the suburbs, but occasionally within the zone of transition if road improvements allow. Indeed, communications have always been important to industry which is why port, waterway, and railway sites have so often been associated with heavy industry using bulky raw materials. These may be nuisance industries distant from housing. The twentieth century, however, has seen most emphasis on road transport, hence **industrial decentralisation** with ribbon development along radial routes and the establishment of industrial/trading estates on ring road intersections. Such suburban locations for light industries producing consumer goods have numerous advantages – motorway and labour access, lower land values and space to expand, the order of a planned layout with flexible power sources, such as gas and electricity, and minimisation of congestion. Consequently, model examples such as the Team Valley Trading Estate in Gateshead have influenced contemporary industrialisation throughout Britain.

Residential districts in the city

The CBD was characterised by the relatively small size of its residential population. This in part reflects the twentieth century trend in most 'western' cities of population migrating outwards from inner districts to suburban zones around their margins. This process is called **residential**

decentralisation and can be expressed and measured in more precise terms by an urban density gradient.

C. Clark (1951) is associated with this model. Rather like the rank size rule, a real life (empirical) regularity was observed and then a formula created to best express it. Clark's **urban density gradient model** is based on actual observation and the analysis of real data rather than being theoretical.

He noted that density of population declined with distance from the city centre.

The resulting model assumed a single centre of employment (like E.W. Burgess and H. Hoyt discussed with C.D. Harris and E.L. Ullman later), and described night-time distribution of population.

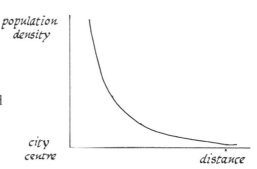

NB: Given that most modern cities have more complex patterns of population density around nuclei of employment – more akin to Harris and Ullman than Burgess and Hoyt – one should perhaps regard the model with a degree of scepticism.

Expressing this mathematically involves the following formula:

$$dx = do\ e^{-bx}$$

where

dx = the population density at distance **x** from the city centre

do = the interpolated (estimated) central population density

e = an exponent of distance (2.718)

b = the density gradient

NB: Although not always stated, assume gross density (the number of people per unit area of all land-use).

Clark recorded that this formula *'appears to be true for all times and all places studied, from 1801 to the present day, and from Los Angeles to Budapest.'*

100 studies all over the world spreading back 150 years have revealed no challenge to the universal applicability of the equation! Yet, just like the rank size rule, attempted explanations have been generally unsatisfactory. They tend to involve the balancing of two desires – access to the city centre for employment and an abundance of living space.

B.J.L. Berry, J.W. Simmonds, and R.J. Tennant (1963), however, suggested rather aptly that *'the poor live near the city centre on expensive land, consuming little of it, and the rich at the periphery consuming much of it. Since the land consumed by each household increases with distance from the city centre, population densities must drop.'*

So what use is the model? It is of limited, if any use, for explaining the form of a particular city. Its real use is in making comparisons between cities and in monitoring changes through time. For example, in 'western' cities the density gradient diminishes as the population of the city increases. Small cities are, therefore, more compact. Larger cities tend to sprawl out as more and more commercial, administrative, and industrial functions compete for space. London's declining density gradient, as the city grew from the early nineteenth century to the mid-twentieth century, illustrates this point.

Also, the density gradient tends to be less steep in younger cities, reflecting more careful planning and the social unacceptability of higher living densities.

City changes can be illustrated by changes in the density curve (see over).

As 'western' cities grew throughout the nineteenth century, central population density increased (lines 1, 2, 3). However, twentieth century CBD expansion and redevelopment of inner city housing sees people exploiting personal mobility afforded by the car and improved public transport, and moving towards the suburbs (lines 4, 5). Clark's model did not predict the resulting **density crater**.

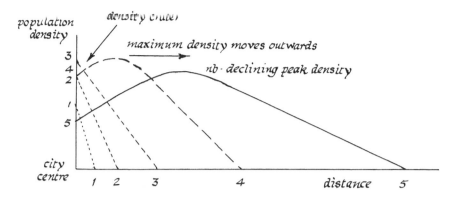

This does not, however, generally work in 'non-western' cities. Indeed, studies of Asian cities have indicated that as the city's population increases so does the central density - while the density gradient remains constant. Hence increased overcrowding and the maintenance of a constant degree of compactness is the norm. The development of a CBD is not usual in 'non-western' cities, although notable exceptions are found where British influence has been marked - such as in Calcutta and Bombay in India, and Nairobi in Kenya.

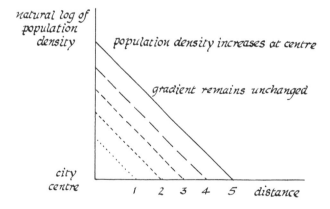

Residential land-use patterns

Residential land-use will normally dominate in a city, accounting for around 45 per cent of the total area. In 'western' cities, certain generalisations may be made about the arrangement of residential land-use.

Within the **CBD** a small residential population may be noted in prestigious dwellings, such as luxury penthouse flats on the highest floors of office buildings or in discrete enclaves such as Mayfair or Belgravia in London.

The fringe area marking the indistinct boundary of the CBD provides the ultimate contrast. This **zone of transition** is characterised by high density nineteenth century housing in various stages of decay and/or renovation. Condemned slum properties may await demolition whilst redevelopment projects (both residential and commercial), workshops, wholesalers, and relic industrial functions complete a landscape of general urban blight.

Immediately beyond the zone of transition are the **inner city** housing districts of unplanned nineteenth century terraced housing, usually without gardens, but increasingly converted to indoor bathroom facilities. These are likely to merge into the **inner suburbs** of large Victorian (often terraced) town houses normally now converted into multi-family dwellings of flats and bedsitters.

Beyond this zone extend linear (ribbon) developments with subsequent infilling of characteristic twentieth century **suburbia.** Whether inter-war semi-detached family houses, or smaller (generally more varied) semi-detached and detached properties in post-war estates, the so-called **urban fringe** offers dramatic environmental and social contrasts, most notably between private and local authority developments.

It could be argued that suburbia is the dominant residential land-use, probably a safe assumption if one regards the space occupied by such low density housing developments as one's main criteria. Suburbia is essentially a product of twentieth century increases in urban population. But it is not simply a question of numbers which explain suburbia - social changes, whereby family sizes have got smaller demanding more dwellings to house the same number of people, are of note. Suburban growth may be interpreted as a reaction against the poor housing conditions of the inner city - effectively a reaction against nineteenth century urban life. Lower suburban land prices and the development of suitable transport systems, plus building society finances, allowed large numbers of people to realise their housing goal of a semi-detached or detached house with gardens in a spacious setting.

Indeed, the role of transport developments are of particular note. Late nineteenth century suburban housing was usually confined to narrow belts adjacent to the newly constructed railways. (The first dormitory settlements dominated by commuters were similarly connected to the main urban area.) After World War I, the spread of the motor car and bus services increased commuter flexibility, further aided by the increasing importance of building societies. This resulted in spreading suburbs no longer constrained by main arterial rail and tram routes. The characteristic extensive areas of low density, single-family housing with ample garden and public open space would reflect public taste of the time. Street layouts and housing styles would vary according to the availability of land, fashion, cost, and tenure - whether private or local authority. Normally residential areas would be characterised by a general uniformity of income and similar social 'classes'. Segregation from other land-uses is normal.

The relationship between wealth, mobility and suburban expansion is of particular note. Wealth gives choice and increased flexibility of movement. One could argue a social parallel with constant outward expansion of suburban housing. Certainly, people with higher incomes have the greatest choice of location and are highly mobile, hence able to choose the most attractive sites on the outskirts, giving rise to distant, so-called **stockbroker belts**. Conversely, people with lower incomes are often found nearer their work in cheaper housing close to industrial areas, whether inner city or suburban. The process of **filtering down** is particularly important whereby, over time, properties filter down to individuals of lower income. As the dwellings become older, and therefore have associated problems, wealthier families move to newer houses further from the town centre. Subdivision of large properties into flats and bedsitters is a common accompaniment. Consequently, the inner city (incorporating inner suburbia) becomes characterised by environmental and social problems - the domain of relatively poor people, such as immigrants, students, the transient, and the aged, collectively least able to pay enough in council tax to maintain essential services. High unemployment, multiple deprivation, ethnic tensions, low educational achievement, high crime rates, and so on thus become synonymous with the inner city.

A final comment should perhaps mention the reverse process to the filtering down just described. **Gentrification** is of particular note in, for example, London. Changing economic fortunes may stimulate the renewal of particularly accessible inner city areas. A former run-down inner city district may consequently become the fashionable habitat of the 'upper middle classes'. Islington and Wapping, in the Borough of Tower Hamlets, are notable examples.

Models of urban land-use

Although every city is unique in its arrangement of streets, buildings, and open spaces, broad similarities in land-use arrangements can often be discerned. Models of urban land-use aim to help describe and explain these patterns. The concentric, sector, multiple nuclei, and British compromise models are of particular interest.

The concentric model (E. W. Burgess, 1925)

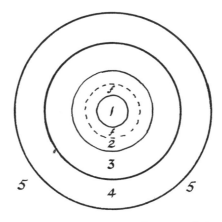

1 loop f factory zone 2 zone in transition
3 zone of working men's homes 4 residential zone
5 commuter zone

Burgess' theory was based on a study of Chicago, USA. His suggestion that urban settlements expand outwards in successive zones from the central core is admirably simple, but has been much criticised. The most usual criticism is that it is anachronistic, with minimal appreciation of the importance of industry or the influence of radial routeways. Clearly without the benefit of hindsight, he should be forgiven his failure to predict increasing personal mobility, allowing residential and industrial decentralisation, dormitory settlements, and out-of-town shopping centres. However, for a concentric pattern to emerge, innumerable radial routeways, each with regularly declining accessibility, would be required. The model underestimates, therefore, the role of transport routes – a criticism specifically addressed by Hoyt.

The sector model (H. Hoyt, 1939)

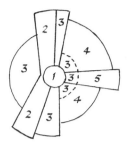

1 *central business district*
2 *wholesale light manufacturing*
3 *low class residential*
4 *medium class residential*
5 *high class residential*
6 *heavy manufacturing*
7 *outlying business*
8 *residential suburb*
9 *industrial suburb*

Hoyt's theory suggests that specific land-uses extend outwards along radial transport routes. Consequently, a series of sectors widen and spread with increasing distance from the city centre. Existing land-use, therefore, persists as more of the same is added to the periphery. Also, mutually exclusive land-uses, such as 'high class' housing and industry, can develop and remain apart. However, like Burgess, Hoyt's theory has been criticised for being dated, inflexible, and underestimating the effect of industry. Again we must accept that Hoyt was writing many decades ago, but acknowledgement of varied rather than predominantly 'high class' housing, along with recognition that some commercial development is likely within residential districts, would have improved the theory.

The multiple nuclei model (C.D. Harris and E.L. Ullman, 1945)

Harris and Ullman's suggestion that some large settlements do not develop necessarily from a single centre allows for far greater flexibility. Progressive integration of a number of separate nuclei to form a mosaic of distinctive land-use zones recognises that whilst some functions welcome clustering, such as legal services, others repel each other, such as convenience retailing. This model was the first to adopt such principles in the explanation of urban structure.

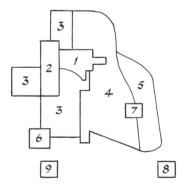

Key as for sector model

Despite these and many other criticisms, it must be said that all the theories provide useful frameworks against which the complexities of reality may be judged. For example, they may provide an interesting introduction to the study of specific cities, especially given that potential accessibility is a common theme and so critical to understanding urban form. However, all three models are based upon American experience, which is why P. Mann's compromise is so welcome.

The British compromise model (P. Mann, 1965)

Mann's model combines arguably the best of Burgess and Hoyt in a more familiar context. Outward expansion from an historic core, yet distinctive sectors grouping like with like may seem socially divisive but is undeniably real. Certainly the familiar theme of east- and west-end contrasts (in acknowledgement of the prevailing winds), not least the recognition of dormitory settlements, makes this model a valuable and interesting framework for urban study.

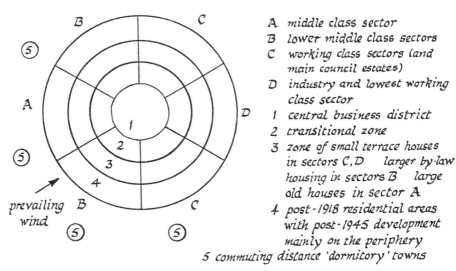

A middle class sector
B lower middle class sectors
C working class sectors (and main council estates)
D industry and lowest working class sector
1 central business district
2 transitional zone
3 zone of small terrace houses in sectors C,D   larger by-law housing in sectors B   large old houses in sector A
4 post-1918 residential areas with post-1945 development mainly on the periphery
5 commuting distance 'dormitory' towns

18

RECREATION AND TOURISM

Over the last 20 to 30 years the demand for leisure, recreation, and tourism opportunities has multiplied in EMDCs. Indeed, tourism is not only one of the fastest growing industries in the world, it is the largest. Recreation and tourism has, therefore, become a major source of employment – a service, labour-intensive *'industry without chimneys'*, with global scope. Recreational time, by definition, is at our leisure – spent passively or actively. Tourism represents use of this time by people *en masse* in locations away from home, whether abroad or not. Geographical concerns tend to be with those recreational and tourist activities that take place out of doors and have environmental implications. We look particularly at the growth of these activities, their impact, and the attempts made to plan for them. Indeed, this is one of the foremost planning issues of our age.

CAUSES OF THE GROWTH IN RECREATION AND TOURISM

1. Growth in population.
2. Greater affluence.
3. Increased leisure time through longer paid holidays and shorter working hours.
4. Inceasing numbers of active elderly people.
5. Improved accessibility and mobility - including the increase in car ownership and affordable charter air flights to overseas resorts.
6. Greater awareness of locations, facilities, and opportunities - through education, advertising, and the media.
7. Convenient access through package holidays, mini-breaks, and so on.

Outdoor recreation is associated with distinct rhythms, with peaks, such as Bank Holidays, presenting problems of road congestion and concentration of activity in limited areas. The volume and frequency of demand is related to the character of the recreational area – its beauty and variety of facilities. Accessibility (in physical or time distance) is also relevant, in that half-day and day trips are generally over short distances. Indeed, there is a clear relationship between the volume and frequency of recreational demand and accessibility measured in either time or distance.

Theoretical pattern of recreational activities around an urban nucleus (M.Dower)

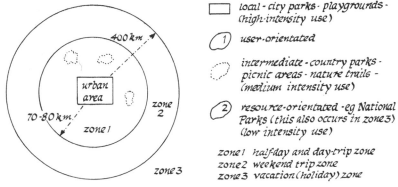

local - city parks · playgrounds · (high-intensity use)

① user-orientated

intermediate - country parks · picnic areas · nature trails - (medium intensity use)

② resource-orientated - eg National Parks (this also occurs in zone 3) (low intensity use)

zone 1 half-day and day-trip zone
zone 2 weekend trip zone
zone 3 vacation (holiday) zone

Although this is a static model,- zones can alter as a result of changes in the transport network, such as motorway construction

After P.J. McBride's (1980) adaptation

Dower's (1970) model of the relationship between frequency of visit and the distance people are prepared to travel is particularly expressive in this context. High intensity use of local facilities could extend, for example, to golf courses on the fringe of the urban area. This **user–orientated** zone has accessibility as its key, and medium intensity attractions, including stately homes and theme parks, are included within its demarcation. The **resource–orientated** zone is more distant, for people are willing to travel much further for the quality of environment associated with, for example, attractive coastlines or National Parks. The intensity of use is, however, reduced. Clearly, the model is static and the zones will vary according to local circumstances, not least increasing mobility facilitated by transport, such as motorway network improvement. Recreational expectations and opportunities continually evolve, hence theme parks have become more numerous, and also used intensively, whilst sightseeing or walking in a National Park is increasingly practised as a day out rather than, for example, camping or guest house based.

Similarly, tourist expectations increase with continuing expansion of the industry. 'Traditional' British destinations are continually re-evaluated, such as from the nearest seaside resort, to the Spanish 'Costas', to Florida and the Caribbean. Each year sees increasing availability of affordable, convenient package holidays to more distant and exotic destinations. It is not surprising, therefore, to see more and more ELDCs promoting their tourist potential despite uncertain benefits to their overall development. Cultural and 'historic' holidays, such as in East Asia, offer much - especially to more affluent middle aged tourists who have long since tired of Mediterranean beaches. Indeed, there is now a remarkably sophisticated market for 'activity' holidays throughout the world catering for every conceivable interest, taste, expectation, and income. East African (photography) safaris represent the longest established illustration of these arguably more stimulating holidays. Their tropical locations virtually eliminate seasonal peaks and troughs by catering for affluent visitors from EMDCs throughout the world.

TOURISM IN AN ELDC: KENYA

Since gaining independence from the British in 1963, Kenya has developed a modest existing tourist trade to its current position as the biggest foreign currency earner. Annually, hundreds of thousands of tourists now enjoy safari and beach holidays accomodated in usually well established and furnished facilities.

The **natural environment** offers much. The **climate** is hot and sunny - except for a summer monsoon affecting coastal areas in April and May. **Beaches**, such as Nyali beach near Mombassa, have long stretches of fine sand, with shark-excluding coral reefs offshore. Safe snorkling and sub-aqua diving is possible, with gaps in the reef used for surfing. However, it is the **wildlife** maintained under government control in Game Reserves and National Parks that attracts most visitors. Nairobi Game Reserve, for example, has lions and cheetahs whilst Tsavo National Park has elephants and black rhinos. Finally, **scenery** including the Great African Rift Valley and mountains such as the extinct volcano Mount Kenya, which is high enough to be glaciated, offer opportunities for climbing, hiking, and studying diverse flora and fauna.

The **cultural environment** is exploited too. Traditional dances, costumes, and craft industries support many villages, but not necessarily to entirely positive effect.

There is much to be commended in Kenya's development of its tourist potential. Game Reserves and National Parks, often on land otherwise virtually uninhabitable due to its low agricultural potential and vulnerability to pests such as the tsetse fly, allow conservation of endangered species previously vulnerable to the ravages of poachers. Infrastructure such as metalled roads, electricity, water, sewerage, and telecommunications has been developed to the benefit of all. Game lodges and hotels provide employment, as do all the supporting services from souvenir manufacturers to airports. Certainly, the income generated by tourism can be invested into developing Kenya's economy, not least much needed education, health care, and family planning programmes.

However, areas unaffected by tourism tend to remain very poor as development is concentrated into areas with tourist potential. Also, the industry is susceptible to the disposable income of people within EMDCs which varies according to the state of their economies. There is a continual 'leakage' of profits out of Kenya to foreign tour operators, airlines, and so on. More difficult to quantify is the discontentment caused within a local population of former hunter-gatherers now performing a degrading pastiche of their traditional lifestyles for affluent visitors often insensitive to their circumstances.

Recreational and tourist demand is not only related to accessibility but also to the nature of the attraction. American tourists, for example, flock to the historic sites of London, arguably regardless of the climate. Domestic recreation can be more flexible in its response to good weather, drawn by the quality of scenery and viewpoints, the ease of car access, parking, toilet, and refreshment facilities – not least, as referred to above, the character of the area. Demand, for example, may be focused into **honeypot** locations such as Tarn Hows in the Lake District or Malham Cove in the Yorkshire Dales. Picnic sites, information centres, gift shops, and readily accessible viewpoints concentrate visitors, arguably preserving relative low intensity use and, therefore, fewer problems in relatively unspoilt areas beyond. Purists might baulk at the commercialised and sanitised character of many honeypots, but they have allowed access to beautiful locations for a far wider spectrum of society, including the disabled. Coasts likewise have to cope with intensive use in that space is both limited and clearly defined, so concentrating the impact of the visitors.

It is possible, however, given the will and finance, to organise, plan facilities, and to manage locations associated with recreational and tourist potential, providing the specific needs of visitors and likely **conflicts of interest** with existing land-uses are understood. Britain is not alone in legislating in order to encourage conservation of areas of natural beauty and their wildlife whilst ensuring public access. Legislation now covers all scales – for example, local Approved Footpaths and Bridleways and Sites of Special Scientific Interest (SSSIs). Local, and at a larger scale, National and Forest Nature Reserves are delimited, as are Areas of Outstanding Natural

Beauty (AONBs) and Heritage Coastlines. However, the National Parks and Access to the Countryside Act (1949) is especially influential because it set up the original ten National Parks in England and Wales defined as '*areas of great natural beauty giving opportunity for open-air recreation, established so that natural beauty can be preserved and enhanced, and so that the enjoyment of the scenery by the public can be promoted.*'

However, even with this history of legislation, protected areas can still suffer from a lack of positive planning, problems of finance, administration, and the perhaps insoluble issue of reconciling conservation with public access. National Parks, for example, are not nationally financed or administered – planning is still primarily in the hands of whichever Local Authorities lie within the Park boundaries. There are still problems of multiple land-use and mutually exclusive land-uses (hence conflicts of interest) within the Parks. Established activities such as farming, forestry, reservoirs, quarrying, and mining, in addition to military training, have legitimate rights – but how are these to be maintained with any sense of harmony given visitor demands? Furthermore, the provision of tourist facilities such as caravan sites and public toilets can blight the very environments to be protected. Clearly, as access to these areas improves and visitor numbers increase, so do the problems of footpath erosion, damage to field boundaries, sheep worrying, forest fires, reservoir pollution, and so on. Social problems also occur. Such areas often suffer high unemployment which is unlikely to be solved by restricting industrial development or encouraging notoriously low paid work in seasonal tourist services. Of particular concern in recent years has been the dramatic increase in sales of housing in National Parks to 'outsiders' for holiday or retirement homes, generating price inflation through planning restricted limited supply and so excluding local purchasers. Most affected are the young, at the foot of the 'housing ladder', many of whom leave and so perpetuate the social consequences associated with rural depopulation, such as closing primary schools, and reduced public transport and (non-tourist) retail services. The degraded social fabric and sense of community within picturesque settlements likened to 'ghost villages' might be difficult to quantify, but is appreciable nonetheless.

Reconciling existing land-use and conservation with public access, therefore, remains one of the most difficult planning issues of our time. But much is being achieved by sensible, informed planning, from landscaping of infrastructure and tourist amenities by, for example, shelter belts of shrubbery and trees, to no longer prohibiting, but harmonising, new buildings – such as modest Housing Association developments – by careful use of local materials and styles traditional to the relevant area. Commercial forest plantations can follow natural contours and be further disguised by irregular boundaries of contrasting deciduous species. Likewise, reservoirs can be multi-functional and quarries landscaped during extraction and reclaimed afterwards. Leisure service occupations need not be part-time, given careful thought to extending visitor seasons with more imaginative appreciation of the environment's potential for 'activity breaks', educational purposes, and so on. Also, years of experience in combating footpath erosion, maintaining agriculturally appropriate access routes, and appreciating the responsible values of most visitors (by, for example, removing all litter bins from honeypot locations) is increasingly evident in many treasured recreational areas.

SELECTED GENERAL READING

As referred to in the *Preface*, it is not the intention of this book to replace, but rather to complement, established textbooks. In keeping with its remit of economy (in the broadest sense of the word) a bibliography has been omitted. However, the selection of general textbooks below offer appropriately comprehensive coverage, exemplification, and illustration.

Bradford, M.G. and Kent, W.A., *Human Geography: Theories and their Applications*, (Oxford University Press, 1977).

Bradford, M.G. and Kent, W.A., *Understanding Human Geography: People and their Changing Environments*, (Oxford University Press, 1993).

Carr, M., *Patterns: Process and Change in Human Geography*, (Macmillan, 1987).

Clowes, A. and Comfort, P., *Process and Landform*, 2nd edn. (Oliver and Boyd, 1987).

Collard, R.A., *The Physical Geography of Landscape*, (Collins Educational, 1988).

Goudie, A., *The Human Impact on the Natural Environment*, 2nd edn. (Blackwell, 1986).

Goudie, A., *The Nature of the Environment*, 2nd edn. (Blackwell, 1989).

Hilton, K., *Process and Pattern in Physical Geography*, 2nd edn. (Unwin Hyman, 1985).

Knapp, B.J., *Systematic Geography*, (Allen and Unwin, 1986).

McBride, P.J., *Human Geography: Principles, Processes and Patterns*, (Blackie, 1980).

Prosser, R., *Human Systems and the Environment*, (Nelson, 1992).

Prosser, R., *Natural Systems and Human Responses*, (Nelson, 1992).

Prosser, R., *Managing Environmental Systems*, (Nelson, 1995).

Waugh, D., *Geography: An Integrated Approach*, 2nd edn. (Nelson, 1995).

Whynne-Hammond, C., *Elements of Human Geography*, 2nd edn. (George Allen and Unwin, 1985).

Wilcock, D.N., *Physical Geography: Flows, Cycles, Systems and Change*, (Blackie, 1988).

INDEX

This comprehensive index has been constructed to ensure fast, efficient use of the book. **Bold** entries indicate terms defined, explained, or discussed in some depth. *Italics* refer to boxed features. An asterisk (★) signifies the use of illustrations.

> **Bold**: Main entry.
> Normal: In context.
> *Italic*: Boxed feature.
> ★: Illustrated.